DATE DUE

DEMCO 38-296

A PRACTICAL INTRODUCTION TO VIDEOHISTORY

A PRACTICAL INTRODUCTION TO VIDEOHISTORY:
The Smithsonian Institution and Alfred P. Sloan Foundation Experiment

Terri A. Schorzman

Editor

KRIEGER PUBLISHING COMPANY
MALABAR, FLORIDA
1993

Riverside Community College
Library
4800 Magnolia Avenue
Riverside, CA 92506
AUG

Original Edition 1993

Printed and Published by
KRIEGER PUBLISHING COMPANY
KRIEGER DRIVE
MALABAR, FLORIDA 32950

Copyright (c) 1993 by The Smithsonian Institution

Library of Congress Cataloging-In-Publication Data

Schorzman, Terri A.
   A practical introduction to videohistory : the Smithsonian
Institution and Alfred P. Sloan Foundation experiment / by Terri A.
Schorzman with David DeVorkin . . . [et al.]. —Original ed.
     p.   cm.
   Presents the results of an experiment conducted from 1986 to 1991
by the Smithsonian Videohistory Program.
   Includes bibliographical references.
   ISBN 0-89464-725-3 (acid-free paper)
   1. Video tapes in historiography. I. Smithsonian Videohistory
Program. II. Alfred P. Sloan Foundation. III. Title.
D16.8.S35855 1993
907′.2—dc20
                                        92-6784
                                        CIP

10 9 8 7 6 5 4 3 2

CONTENTS

PREFACE

Videohistory[1] is the video recording of visual information as primary historical evidence and involves a historian in shaping the original inquiry. From 1986 to 1992, historians at the Smithsonian Institution (SI) examined videohistory's effectiveness as a tool for historical research. The Alfred P. Sloan Foundation, of New York, convinced of the potential value of videohistory, funded the five-year experiment, known as the Smithsonian Videohistory Program (SVP). The SI established an advisory committee and program staff, and invited its institutional scholars to propose a variety of videohistory projects in the history of science and technology that they anticipated would benefit existing research. The committee oversaw policy, awarded funding, and evaluated completed projects; the staff coordinated and managed individual projects as well as provided daily guidance for the overall program. Historians and academic consultants also reviewed the videohistory materials. The experiment was intended to work out new guidelines and methodologies for helping historians interested in using video technology in their research.

By the end of 1991, eighteen historians had created over two hundred and fifty hours of tape on twenty-two different subjects. The tapes, transcripts, finding aids, and supporting materials, including reviews, tape logs, production notes, and reports, have been deposited in the Smithsonian Institution Archives, and are open to research. Historians, professors, scientists, film and television producers, and exhibit designers have already used the collection for a variety of purposes. Federal agencies and private industry have also requested copies of tapes and transcripts relevant to their interests.

The Smithsonian Videohistory Program set out to determine where videohistory—as an extension of audiotaped oral history—is best used, a necessary question because of the time, expense, and complexity of videotape recording. The SVP determined when videohistory was an appropriate tool for historical research and then determined what recording approach would be most effective based on the nature of individual projects. They asked: Is videohistory most effective when recording people interacting with other people, with objects, with a process, or within an environment? Is it valuable in portraiture, particularly when speech and mannerisms reveal an individual's personality? Does the lack of strong visual evidence undermine the effectiveness as a "talking head" interview? Are videotaped interviews ineffective when reflecting upon abstract subjects or past events? Is video more effective in capturing emerging technology and contemporary work patterns?

This book reviews video projects and techniques used prior to the SI experiment, recounts the creation and management of the SVP, and establishes the conditions under which videohistory projects were developed, processed, and evaluated. Chapter One includes a brief history of the program, and addresses its conditions and assumptions. Chapter Two focuses on how to create visual evidence in historical documentation with examples from SI projects. Chapter Three explores technical requirements for producing videohistory and how the SI addresses special needs for storing and preserving videotape and transcripts. In Chapter Four, four historians, a director, and a scholar from the field of communications discuss their experiences and provide impressions of the value of videohistory as a research tool. The book concludes with an appendix of sample forms and supporting documentation, and a summary of each videohistory project.

NOTES

1. The word *videohistory* has been used since the late 1960s to refer to videotaped interviews that become historical records, the value of which might not be known for decades. It was spelled as *video-history*, then *video history*; the SVP adopted *videohistory*.

ACKNOWLEDGMENTS

This book—and Videohistory Program—would not have been possible without Arthur Singer and The Alfred P. Sloan Foundation. Mr. Singer recognized the possible uses of video in historical research and brought the Foundation and Smithsonian together to begin the task. We thank the Foundation and Arthur Singer for their generous support.

The Videohistory Program was the combined effort of many, many people. Among those people were Smithsonian professional and support staff, over 300 participants, and technical crews and video directors from around the country. Special thanks go to the eighteen historians who were willing to undertake such experimental research. Numerous corporations and agencies, archives, laboratories, local and national museums, and educational institutions also contributed to our work. Without the support of such establishments, many projects would never have been completed.

The Smithsonian Institution Archives set aside space for the videohistory operation and provided archival and administrative support. In particular, Pamela Henson, one of the guiding forces behind the creation of videohistory at the Smithsonian as well as one of its most thoughtful practitioners, led us through the world of archival processing and helped formulate concepts when exploring notions of the "visual dimension."

The Space History Division of the National Air & Space Museum also provided an extraordinary amount of administrative support. We owe thanks to Mandy Young for helping us create a workable and well-organized administrative operation.

The Videohistory Advisory Committee, David DeVorkin, Marc Pachter, Nate Reingold, and William Moss, offered consistently positive input for the duration of the program, especially in the early years when formulating policy and procedures and selecting projects were critical. Special thanks go to Bill and David for their editorial and personal support. It meant a lot.

Finally, I can't begin to thank the Videohistory staff, Joni Mathys, Phillip Seitz, and Alex Magoun, enough for their incredible work, remarkable stamina, creativity, and intellectual support. In our five years together, we saw the completion of two masters degrees, the successful passing of doctoral exams, and the birth of a baby; we shared travel adventures to the Soviet Union, Panama, New Zealand, and points in between, ate more Belgian chocolate and "sour and cream and onion" potato chips than I care to remember, and became versed in the language of science and technology as well as in the perplexing world of federal bureaucracy. To them, I am grateful and I'll treasure the memories. I also thank Joni for becoming one of my best friends. She gave much of herself to this program, which became a labor of love for both of us.

CONTRIBUTORS

Editor and primary author: Terri A. Schorzman, Program Manager, Smithsonian Videohistory Program, Smithsonian Institution. M.A., University of California, Santa Barbara.

Contributing authors:

David DeVorkin, Curator in the History of Astronomy and Space Sciences, National Air and Space Museum, Smithsonian Institution. Ph.D., Yale University.

Stanley Goldberg, Historian of Science, Washington, D.C. Ph.D, Harvard University.

Kerric Harvey, Assistant Professor, George Washington University. Ph.D., University of Washington.

Steven Lubar, Curator of Engineering and Industry, Smithsonian Institution. Ph.D., University of Chicago.

William Moss, Director, Smithsonian Institution Archives, Smithsonian Institution. M.A., Columbia University.

Carlene Stephens, Curator of Engineering and Industry, Smithsonian Institution. M.A., University of Delaware.

Brien Williams, Media Producer, Washington, D.C. Ph.D., Northwestern University.

1
ORIGIN AND COURSE OF THE SMITHSONIAN VIDEOHISTORY PROGRAM

INTRODUCTION

Beginning in 1986, historians at the Smithsonian Institution (SI) participated in a four-year experiment in using video to document American twentieth century science and technology. The work was later extended for a fifth year. With funding from the Alfred P. Sloan Foundation, and with guidance and technical support from the Smithsonian Videohistory Program (SVP), a special office created for the purpose, these historians created a body of visual evidence, such as documentation of artifacts, environments, and group interactions, that supplemented traditional historical documentation.

ANTECEDENTS TO THE SI EXPERIMENT: VIDEOHISTORY, 1960–1985

The notion of using motion picture cameras caught the imagination of some historians in the 1960s. They saw the apparent benefits of adding a visual dimension to their research, and encouraged their fellow historians to experiment with new technology. Their enthusiastic conclusions, however, were not derived from systematic testing. Nevertheless, their efforts and products laid a groundwork of trial and error for later work in videohistory, and formed the background of experience against which the Smithsonian experiment was conducted.

In 1967 Louis Shores spoke to the First National Colloquium on Oral History about the emergence of a new dimension, the visual dimension, in the collection of oral history materials. Shores asserted that technology had delivered the opportunity to capture voice, movement, and presence. He suggested that certain subjects warranted the visual dimension to strengthen the audiotaped interview, particularly for those who demonstrated their art or interacted with their environment in unique ways. "I am suggesting that we fortify oral history with visual history, and capitalize on developments in the audio-visual movement."[1]

Joe B. Frantz of the University of Texas touted the power of visual presence when he spoke about the university's program to videotape notable academic historians in the classroom. From 1962 until 1964, with Ford Foundation support, the university videotaped forty-one lecturers in the fields of their specialties. According to Frantz, the project helped the university capture the presence of people that future generations of historians would find instrumental to their understanding of the profession. Sarah Diament of Cornell University urged historians at the Fourth National Colloquium on Oral History to adopt the visual medium to conquer their fear of technology, to master new techniques, to record movement, and above all, to venture beyond stereotypical historical research in the acquisition of new methods. She believed that film and video could be used effectively as historical research tools, not just as entertainment, and maintained that "the most exciting historical research in primary source materials involves the historian in just such an immediate experience."[2]

People in the medical field also accepted the value of visual evidence when they videotaped autobiographical interviews of well-known physicians for use in medical education. In 1966, David Seegal, an adjunct professor for medical education at Columbia University, thought that the method "captured the whole presence" of leaders in the medical field and that videotaped interviews with medical scientists and teachers both documented medical history and inspired students of medicine.[3]

By the 1970s and early 1980s historians and their organizations videotaped oral history interviews as a regular component of their work. Their work provided greater cumulative experience across a wider range of subjects and situations. Conclusions based on this experience became less assertive and increasingly reflective.

For example, in 1977, Gallaudet University in Washington, D.C., one of the nation's leading education centers for the deaf, expanded the visual dimension of oral history when it used videotape to record interviews with both those who are deaf and with hearing people who work with the deaf. Interviews were conducted in international sign language by Gallaudet personnel. This use of the technology allowed the university to record its history through visual communication and established Gallaudet as an important producer and user of videotaped interviews.[4]

Beginning in the early 1980s, James Briggs Murray, director of the Schomburg Center of New York Public Library, added video to its oral history interviews arguing that "so much more of a person can be perceived from seeing moods, expressions on a screen than can possibly be picked up from simply a voice recording or cold black letters on a white transcript. . . . " He found that the emotions and movements of interviewees brought viewers "about as close as we and future generations might ever expect to get to them long after they have left our physical midst." Librarian F. Gerald Handfield found that videohistory gave a more accurate message by conveying the impact of body language and that it allowed the interviewee to immediately refer to an object or site. Journalist W. Richard Whittaker felt that video added "depth and breadth to recollected experience." Recording body language and visual reactions were also important to Whittaker. Other researchers and scholars noted similar experiences with videotaped interviews, yet they cautioned that an interviewer must think seriously about visual content before beginning videohistory. If done correctly, filmmakers Elizabeth Jameson and David Lenfest asserted, the product will make the oral history more "complex, more moving, and more undeniable," a goal for those who wanted to use oral and visual sources for public education.[5]

By the mid-1980s, oral historians began to grapple with some of the problems as well as reflect on the merits of videohistory. In 1984, Joel Gardner, then assistant director of the Louisiana Division of the Arts, worried that professional video producers had begun to co-opt the use of video in the humanities. "Form suddenly holds priority over content; a slick, pretty production usually wins out over a solid one that is not as well produced." Gardner urged professional historians to reclaim the field since they were committed to "proper representation and analysis of interviews" in specific fields.[6] On balance, some filmmakers attempted to protect the integrity of their sources by preserving out-takes, and by including information in a production that was true to the style and personality of an interviewee. Ken Burns, film producer, spoke to the 1987 annual meeting of the Oral History Association, in St. Paul, Minnesota, about his attempt to do so with his work.[7]

Some projects began to use video as an adjunct to oral history interviews. The University of California, Los Angeles, tried the technique to a limited extent, and Yale University's program in Oral History, American Music added video to its activities. According to Vivian Perliss, the Yale program included video because,

> . . . if the aim of oral history is the preservation of the personality of an interviewee as well as what he has to say . . . then the use of video tech-

nology is as important as the use of the tape recorder was when it first became available. Our use of the video interview is usually with the composer who had already been interviewed on audiotape. It is considered an adjunct to the oral material just as spoken interviews can add to written documentation.[8]

Tom Charlton of Baylor University noted several concerns about the use of video in recording oral history, including equipment incompatibility and rapidly changing formats, the instability of video for long-term preservation, and the intrusiveness of cameras, lights, and technicians during an interview. Even by 1990, some oral historians were still discouraged on the latter point, and saw the choice to use video as a complex one. They were concerned that video would intrude on the privacy of an oral history ("talking head") interview. Charlton, however, encouraged oral historians to become "competent and professional gatherers and preservers of the recollected past, hoping that technology will support and enhance, not hinder, their endeavors."[9]

Among those who used technology to enhance research endeavors were local, state, and federal agencies, foundations, historical societies, and universities. These institutions recorded both elites and non-elites, experimented with one-on-one and group interviews, and recorded in studios and at field locations.[10] While a consensus on the definitive use of video in oral history had not been reached, these efforts certainly encouraged further experimentation. Thus, by the mid-1980s, a variety of projects, such as those conducted by the Alfred P. Sloan Foundation on science and national policy, enriched the background of accumulated practice against which the Smithsonian built its program and measured its results.

ALFRED P. SLOAN FOUNDATION'S INTEREST IN VIDEOHISTORY

By 1985, when the Alfred P. Sloan Foundation sought out the Smithsonian as a possible home for a national videohistory center, the Foundation already had videohistory experience, thanks to Arthur Singer, Foundation vice president. In 1966, Singer, then at the Carnegie Corporation, and journalist Stephen White, had suggested that the Carnegie conduct a video recorded conversation with Robert Oppenheimer and his colleagues about building the atomic bomb. They hoped to create an informal biographical portrait of Oppenheimer by interviewing him in his home or office—as an ordinary conversation organized around one man and his career. Friends and associates would gather at different times to discuss those parts of Oppenheimer's career with which they were familiar. The Carnegie Corporation's president did not approve the proposal and the opportunity was lost. Within four months, in February 1967, Oppenheimer was dead.

Fifteen years later, Singer, then at the Sloan Foundation, resurrected the idea of videotaped interviews. Al Rees, Sloan's president, liked the idea of "archival television" and agreed that its objective would be to add a new dimension to historical research materials. He and others were concerned that television technology had been directed almost entirely toward entertainment, even when the intent had been to produce a documentary. "Some educational activity has taken place, but for the most part on budgets so low that the potentialities of the technology have necessarily been left unexplored."[11]

The Foundation appropriated $16,000 in 1981 for the effort and supported well-focused projects in the fields of science and public policy. They brought together moderators, participants, and crews to complete each project. The first taping took place in 1981 at the Massachusetts Institute of Technology (MIT), with scientists, engineers, and economists. More than a dozen people were interviewed four or five at a time about their role in the Project Charles study group in 1949 and 1950. Project Charles, organized in 1949 in response to the Soviet's detonation of a nuclear device, made recommendations for civil defense.

The Sloan Foundation conducted other group interviews and by 1984 had appropriated $100,000.00 for the creation of videohistory projects, both at the Foundation and elsewhere. These projects addressed President Truman's decision to accelerate the development of the hydrogen bomb, the Cuban Missile Crisis of 1962, the service of former U.S. ambassadors to the Soviet Union, and the Test Ban Treaty of 1963. Other institutions, with Sloan support, also developed videohistory materials. For example, MIT conducted a three-year project on the development of the digital computer; Brandeis University taped the recollections and experiences of I. I. Rabi, physicist and Nobel laureate; Tufts University recorded recollections of former senior officers of the Department of State; and the Gerald R. Ford Foundation, in collaboration with other presidential libraries, taped a group conversation of former first ladies.[12] The Sloan Foundation houses and distributes the tapes.

By the close of 1984, the Foundation reported that the results "clearly demonstrated the value of this technique for both historical and instructional purposes." Subsequently, Foundation sponsors thought that videohistory needed an institutional home that would continue the experiment as a formal program.[13]

THE SMITHSONIAN RESPONSE: AN EXPERIMENTAL PROJECT

Singer, buoyed by the success of these Sloan-supported projects, approached Robert McC. Adams, Smithsonian secretary, in June 1985 with an offer of substantial funding. He inquired if the SI might be interested, with Sloan funding, in capitalizing on the videohistory experience to date. Adams distributed Sloan Foundation supporting materials to archivists, curators, and historians at the Smithsonian to assess their interest in, and the potential effectiveness of, such a program at the Institution. They reviewed the materials and agreed that adding a video component to collecting historical evidence at the SI seemed promising. They were unconvinced, however, that the addition of the visual element in historical research had been tested sufficiently to justify a major long-term commitment, and were skeptical of earlier claims of self-evident significance. They felt that it had not been examined systematically and critically enough to warrant substantial investment in what promised to be an expensive undertaking and were concerned with methodology, technology, and programmatic implications.

For example, Nathan Reingold, then editor of the Joseph Henry Papers at the SI, was wary of establishing an independent, non-thematic, research program and insisted that committing to a videohistory project be part of the Institutions's long-term obligation to gather source materials in the history of science. Reingold stated that a trained scholar, well-versed in a subject area, should conduct videohistory interviews, rather than a moderator who simply guides a conversation of peers. "We should not allow the taping session to be a free-form discussion by elderly participants, but carefully introduce an analytical structure. . . . Unless we put in the context and framework, the videotapes will have limited validity both as historical sources and as teaching tools."[14]

Roger Kennedy, director of the National Museum of American History (NMAH), acknowledged his interest in videohistory as a means of both storing and disseminating information. He cautioned, however, that research ought to drive the technology, rather than the other way around, and that videohistory projects should be composed of "systematic inquiries rooted in the best social history."[15] William Moss, director of the Smithsonian Institution Archives (SIA) agreed with Sloan reports that video can capture the excitement of collegial discussion and that it "adds a dimension that is not gained quite so well in any other way." Yet, he urged against the creation of a "video history center" that sought out projects to use its equipment; he would rather see the methodology of videohistory made available to projects where the medium would support integrated research. "The device does not stand alone. . . . moreover, it has limitations not addressed in the Sloan prospectus." Moss noted those limitations as the reluctance of a group to approach delicate areas, to favor dominant personalities, and to speak in a familiar shorthand that omits details and connecting logic necessary to the future researcher.[16]

Adams found Singer responsive to these concerns and assured him that the Foundation would not interfere with SI objectives. Adams expressed interest in the SI becoming the leader in the field of videohistory and formed a committee to prepare a formal proposal. David DeVorkin of the National Air and Space Museum (NASM), John Fleckner of the National Museum of American History (NMAH), William Moss of the SIA, Marc Pachter of the National Portrait Gallery (NPG), and Reingold, formed the initial committee.

The group chose to avoid the constraints of broadcast television and documentary productions. In doing so, they hoped to show to SI management that the collection of historical visual evidence was a worthy investment of time and money. Throughout the Program they encouraged that historians use several recording formats and styles in a variety of situations, examined the results closely, and applied a systematic review of completed projects. Their primary purpose was to discover if videohistory was a workable research tool, and to develop a set of principles for appraising the evidential value in its applications. Reviewers included the principal investigators themselves, the Program's advisory committee, scholars from around the country, and technical experts.

The committee's first goal was to establish criteria to guide design and development of the initial proposal and, later, the program itself. The group, acting upon Reingold's suggestion, selected the theme "Science in National Life." Science was one of the Institution's long-standing areas of interest, in addition to artifact collection, research, and education. Second, they insisted that a videohistory project must preserve and ensure the availability of materials for scholarly use, and they encouraged evaluation of videotapes by scholars as a means for constructive feedback. Third, they insisted that SI ongoing historical research, not video technology, drive the program, and that results should add a significant new documentary component to the field of study. Fourth, they urged experimentation with a wide range of setups, from one-on-one interviews to more complex multicamera group interviews. Finally, they addressed several points necessary for implementing a program, including the need for proper archival preservation and documentation, for the cooperation of outside specialists, both intellectual and technical, for developing a relationship with other SI archival and technical facilities, and for deciding upon the general organizational structure of the project.[17]

For the next several months, the committee, along with other interested SI staff, met regularly to create a formal proposal that would, according to Adams, reflect a legitimate academic purpose for the use of videohistory. To the committee, this meant that specific video projects must augment ongoing historical research at the SI, and they agreed to pursue the grant for this reason alone. The final proposal was submitted in mid-May for consideration at the Foundation's board meeting in June.[18]

The proposal addressed both philosophical and structural issues, and stressed the group's interest in pursuing new forms of research. The proposal outlined the theme "science in national life" by placing the program in the broader SI agenda:

> The Smithsonian's programs in historical research and documentation of contemporary science aim to record and elucidate the complex relationship of science to contemporary society. Although these programs may study different elements of the recent history of science and from different perspectives, they are all guided by a strong commitment to American cultural and social history. This common basis inspires the over-all theme we have chosen to link the many on-going and proposed Smithsonian programs in the history of contemporary science: "Science in National Life."[19]

The proposal also stated that the program must be an active archival effort to preserve all documentation generated by the program including project files, research and planning notes, copies of videotapes, audiotapes, and supplementary materials. Proper procedures for the preservation and control of documentation derived from videohistory projects would ensure access to materials for scholarly use.

The Sloan Foundation approved the proposal and awarded the Smithsonian a four-year million dollar grant on June 17, 1986. Granting guidelines stated that funds would be awarded in installments of $100,000.00 the first year and then $300,000.00 per year thereafter.

PROGRAM DESIGN AND DEVELOPMENT: PUTTING IT IN PLACE

Planning proceeded throughout the summer of 1986. The formal advisory committee, consisting of DeVorkin as chairman, Moss, Pachter, and Reingold, and with the advice and assistance of Pamela Henson, SIA historian, and Amanda Young, NASM administrative officer, began putting the program, then called the Sloan Videohistory Project, in place. They established a first-year budget, found office space and furniture, and announced the positions of program manager and assistant. They wanted an office that would administer the overall program and coordinate, manage, and support individual video projects.

They required that the program manager balance office administration, archival management, and technical coordination with an "intellectual partnership" with each researcher. This, they assumed, would "ensure that a systematic study was carried out on the value of videohistory as a mode of historical documentation."[20] The program assistant would provide administrative, archival, and research support. The positions were announced nationwide through professional and academic vehicles, and candidates were interviewed late in the year. Terri A. Schorzman was selected as program manager and Phillip R. Seitz as program assistant. They began work in early 1987.

On September 30, 1986, the program sponsored a seminar to review production and use of videohistory in historical research. The committee invited scholars who had developed video documentation projects to present their work to SI historians and to discuss the process by which video sessions were conceived, designed, executed, and used for historical research. Those attending the seminar listened to advice on formulating projects and applying video techniques to their work.

Based on their recent experience, presenters Jack S. Goldstein of Brandeis University, Jeffrey Sturchio of the American Chemical Society, Robert O. Johnson of the Center for History of Chemistry, and Charles Weiner of MIT addressed issues that were of concern to SI historians. Goldstein emphasized the importance of technical competence of a video crew, of having group participants who knew one another, and designing effective multi-participant sessions. Sturchio and Johnson stressed that video should be used as an adjunct to research: to capture something not available in any other way; to capture unique events as they happened; and to reenact experiments or techniques that make artifacts "come alive." They also suggested that video record obsolete crafts and technologies where creators could take apart machinery and instruments to show how they were constructed and designed, and to assess the problems and compromises during creation.[21]

Weiner pleaded for experimentation using the visual medium and suggested that the program explore various documentation techniques, including that of "visual ethnography," i.e., recording events as they happen with an educated instinct that those events will have historical significance. He showed examples from his video work when he recorded the Cambridge, Massachusetts, city council hearings on recombinant DNA research. As a result of his effort, he emphasized the importance of the historian as witness—and that the historian must dare to anticipate what might be useful documentation for the future. Historians must, however, be prepared to follow the story through to the end since long-term commitment to both the project and the preservation of material is critical. Weiner cautioned that historians not become "videohistorians," but remain scholars who use video as one means of documentation.[22]

Weiner's presentation encouraged discussion on several critical points, including selectivity and the nature of taping contemporary events. SI historians and the committee decided that there would always be practical limitations of what could be recorded. And, since the program planned to support only ongoing SI research in the history of science and technology, projects would be in areas in which they were most familiar. The group also considered the possible overlap between collecting information on current events and retrospective accounts, and noted that Weiner's methods were the closest to investigative journalism. While seminar participants anticipated that some projects might respond to contemporary events, they stressed that their projects were based in historical inquiry and that they would balance the two documentary approaches.[23]

PROGRAM STRUCTURE:
A FORMAL OPERATION

Following the seminar, the committee and historians began the first of several meetings that were concerned with analysis and evaluation of the video process, issues that emerged from the seminar.

The committee decided to establish a central operation with formal guidelines and a review process. This comprehensive perspective allowed the central staff to develop standards, procedures, and reporting deadlines while building on the cumulative experience of each videohistory project. The central operation gave individual historians a support process that provided the best opportunity to achieve their objectives without having to start from scratch or attend to administrative and archival details; it also allowed them to maintain standards for selection, preparation, production, assessment, and preservation of videohistory materials.

The committee, as an advisory body, was interested in theory, policy, and procedure. In general, they provided guidance on planning and policy implementation, and on establishing administrative procedures. They also selected potential video projects. With that in mind, William Moss wrote to the committee stating that he was wary that staff and historians might be pulled toward video production for broadcasts and exhibits, rather than undertake the intended experimental, archival approach. He encouraged the committee to concentrate on unknown aspects of videohistory, and to fund projects weighted towards original inquiry and archival research. The group agreed.[24]

Thus, during the first two years of the Program, we focused on taping projects, on refining techniques, and on testing for research value. At the close of the second year we invited projects that documented the history of technology for further assessment of the quality of moving visual data. We assumed that by recording unique or dying technologies and techniques, or by documenting diverse contemporary manufacturing processes and working environments, visual images through the use of artifacts, diagrams, photographs, equipment, and the environment might prove to be the strongest and most intellectually stimulating component of video projects beyond what was traditionally available. Third and fourth year projects provided such images for analysis.

We, the staff of the SVP, created a formal application packet and established review procedures for project proposals. The packet consisted of guidelines for application and outlined the roles of the historian (as principal investigator), the committee and program office, and outside technical contractors. The guidelines were very clear. Historians were told that projects would be selected on how well the proposal reflected the theme "Science in National Life" and whether or not it had a clearly defined purpose. In addition, we asked that historians develop project proposals that demonstrated how and why video would enhance their research; that sought the collection of new evidence; and that outlined possible research questions. Proposals also required that the historian recommend an outside scholar to evaluate the product.

Roles for historians, staff, and technical advisors were also clearly stated to ensure that everyone was aware of his or her responsibility, as well as the responsibilities of others, to complete a successful video-history project. Historians worked closely with staff and technical advisors in all phases of planning and budgeting, were responsible for the intellectual content of their project, made all preliminary contacts with participants and site representatives (if taping occurred on-site), selected taping locations, scheduled dates for taping, and determined which objects would be included in the interview. Upon completion of a video shoot, they were asked to prepare name and word lists to assist transcription, to review transcribed documents, and to prepare a report based on their observations and conclusions about the videotaping process. Response to these requests were not always forthcoming.

We facilitated all phases of planning and production, and helped historians determine what kind of visual information might add to the historian's existing research. We also oversaw the budget, took care of logistical arrangements, hired video producers and crews, prepared contracts, handled all travel and other paperwork for each project, and occasionally provided research services for historians. We juggled the progress of projects, budget, production and administrative issues, with personalities, problems, and technical arrangements for several simultaneous projects. When Joan Mathys joined the SVP in 1988, she performed most archival processing duties. She arranged for tape duplication and transcription; audit-checked, edited and prepared final transcripts, which included abstracts of important visual information and time code indexing; and wrote supporting project documentation such as finding aids, indices, and catalogs. Alexander B. Magoun joined the staff part-time in 1989 to assist with archival processing. Andrew Szanton, Laura Kreiss, Maureen Fern, and Pablo Jusem also provided temporary archival and research assistance at various times in the program.

The role of technical advisor was also defined in the application packet. In consultation with Program staff and historians, the advisor—as producer/director—hired a crew and equipment, planned technical and logistical requirements, prepared set design (camera placement, lighting, sound), and assured that production was completed in accordance with plans. After the shoot, he or she prepared an indepth report about the production, which highlighted technical concerns and addressed issues that provided commentary for project assessment. This created a body of knowledge that the Program used when planning subsequent projects. Reports addressed a variety of issues, such as a crew's adaptability to the videohistory style of taping, or noted how to make participants comfortable during the process. For example, Selma Thomas, producer of the twentieth century small arms designers video project, confirmed the importance of a one camera shoot to keep the interview process cordial and flexible. She noted that the decision to use one camera was important in creating a personal, less intimidating environment for the interviewee. "Behind the [interviewee's] coat and tie is a thoughtful, at time frustrated and at times triumphant, designer. . . . What emerges is a thoughtful record."[25] Her report, noting reasons for use of one camera, helped us plan for similar projects.

Program guidelines also required an evaluation process to review completed projects.[26] These reviews assessed the usefulness of video as a research tool and were conducted by a variety of people. First, historians themselves reflected on their experiences and either met with the committee to discuss the outcome of their projects or wrote reports analyzing their work. Second, committee members reviewed projects and wrote critiques; third, outside scholars, selected by the historians, evaluated the product for what video may or may not have added to historical documentation in a particular subject area. Finally, technical advisors also reviewed footage for quality and post-production possibilities. Essays in Chapters Three and Four address results of the evaluation process.

PROJECT RESULTS

By the close of the SVP in July 1992, eighteen historians had created over 270 hours of tape on twenty-two different subjects. Topics included instrument building in the space sciences, the development of the Manhattan Project, the emergence of robotics technology, a study of computer hardware and software design, past efforts and the current status of conserving endangered species, and changing industrial technology of two century-old New England companies.[27] Costs averaged $1,500 per day for a single camera crew and upwards of $5,000 per day for a multicamera studio shoot. Individual projects, organized by session (either a single session or many), took at least one year to complete, including all phases of planning, production, and archival processing. In some cases, projects consisting of multiple sessions took several years before they were finished and opened for public use. Videotapes, transcripts, and supporting materials are deposited in the Smithsonian Institution Archives.

CONCLUSION OF THE SVP

The Smithsonian Videohistory Program created a body of historical evidence that contributed to historical research projects at the Smithsonian Institution. The SVP added data to the archival resources at the Smithsonian Institution Archives, provided material for exhibitions and education, and developed a body of experience from which others may draw conclusions about the value of using video in historical research. The material has been analyzed, numerous requests for copies of videotape have been received, and results received eager reception when shared in national and international forums. The materials will continue to benefit scholars, the field of history, and the methodology of videohistory. We hope that our experience will aid other research projects.

We did not succeed, however, in sustaining videohistory as a pan-Institutional research support service. The generous gift from the Alfred P. Sloan Foundation did not produce a continual existence envisioned by both the Foundation and by early supporters from the SI administration. Sloan's initiative provided seed money; we hoped that it would attract financial support from the Smithsonian and others to establish a permanent videohistory presence so that significant projects could rely upon the expertise generated in the experiment. Although individual scholars at the SI remain enthusiastic about the contribution that videohistory techniques can make to their research, and while they intend to seek support for projects through various funding sources, neither the Smithsonian itself nor other fund-granting agencies continued the program as a pan-SI research service.

The general economic climate and shrinkage of the Smithsonian's own funding base partially contributed to the end of the SVP. Also, the need for distributable products, as opposed to pure research projects (without a definite use), became increasingly more desirable. Finally, none of the projects undertaken were sufficiently complementary to the SI's current interest in biological and cultural diversity and public education to capture the dedication of management. SVP projects were not alien to these themes, and some clearly were compatible and contributory, but were not easily used in the public service of these themes. Without these compelling attractions, as sound as the products may be in terms of historical evidence, the videohistory service could not compete with more exciting alternatives.

We hope that inherent, cumulative long-term benefits will determine the true success of the Program. When everyone is doing videohistory as a matter of course rather than as an experimental venture, the Smithsonian Videohistory Program will be valued as one of the progenitors that established the field.

NOTES

1. Louis Shores, "Directions for Oral History," *Oral History At Arrowhead: The Proceedings of the First National Colloquium on Oral History*, (1967), 53–66. For analyses of the use of film and video in folklore and anthropology, see Bruce Jackson, *Fieldwork* (Urbana, 1987); and John Collier, Jr., and Malcom Collier, *Visual Anthropology: Photography as a Research Method* (New Mexico, 1986).
2. Joe B. Frantz, "Videotaping Notable Historians," *The Third National Colloquium on Oral History*, ed. Gould P. Colman (New York, 1969), 89–101; Sarah E. Diament, "Can Film Complement Oral History Interviews?" *The Fourth National Colloquium on Oral History*, ed. Gould P. Colman (New York, 1970), 122–27.
3. David Seegal, "Videotaped Autobiographical Interviews," *Journal of the American Medical Association*, 195 (Feb. 21, 1966), 138–40.
4. "Oral History of and for the Deaf Begins at Gallaudet," *Oral History Association Newsletter* (Fall 1977). See also, David L. Wilson, "Governors Past: A Video History Project," *Journal of Instructional Media*, 6 (1978–79), pp. 253–264, for discussion of project that videotaped Florida's past governors; and Don Page, "A Visual Dimension to Oral History," *Journal* 2 (Canadian Oral History Association, 1976–77), pp. 20–23 for an analysis of what video can contribute beyond audio techniques as well as an identification of potential problems with videotaped interviews.
5. James Briggs Murray, "Oral History/Video Documentation at the Schomburg Center, More Than Just 'Talking Heads'," *Film Library Quarterly*, 15 (no. 4, 1982), pp. 23–27; F. Gerald Handfield, Jr., "The Importance of Video History in Libraries," *Drexel University Quarterly*, 15 (Oct. 1979), pp. 29–34; W. Richard Whittaker, "Why Not Try Videotaping Oral History," *Oral History Review*, 9 (19810), pp. 115–124; Elizabeth Jameson and David Lenfest, "From Oral to Visual: A First-Timer's Introduction to Media Production," *Frontiers*, VII (1983), pp. 25–31. See also, Ron Chepesiuk and Ann Y. Evans, "Videotaping History: The Winthrop College Archives' Experience," *American Archivist*, 48 (no. 1, Winter 1985), pp. 65–68.
6. Joel Gardner, "Oral History and Video in Theory and Practice," *Oral History Review* 12 (1984), pp. 105–111. In the same article, Gardner defined the difference between complementary and supplementary interviews. Complementary interviews reflect aspects related to but not covered in an audio interview; supplementary interviews are recorded after the interview is complete to make use of information for a video production, or to document a single visual aspect. For further discussion on importance of the role of historian in making documentaries, see,

Daniel J. Walkowitz, "Visual History: The Craft of the Historian-Film-maker," *The Public Historian* 7 (no. 1, Winter 1985), pp. 53–64; and Gerald H. Herman, "Media and History," *The Craft of Public History: An Annotated Select Bibliography*, ed. David F. Trask and Robert W. Pomeroy III, (Westport, CT, 1983), pp. 313–350, for introduction to use of visual sources and full bibliography.

7. See also, Pamela M. Henson and Terri A. Schorzman, "Videohistory: Focusing on the American Past," *Journal of American History*, vol. 78, no. 2 (Sept. 1991), pp. 618–27, for brief discussion of other historical documentaries, producers, and use of archival footage.

8. Vivian Perliss, "Oral History as Biography," in *The Past Meets the Present: Essays on Oral History*, ed. by David Stricklin and Rebecca Sharpless (Lanham: University Press of America, 1988), pp. 43–56.

9. Thomas L. Charlton, "Videotaped Oral Histories: Problems and Prospects," *American Archivist*, 47 (Summer 1984), 228–236; see also Eva M. McMahan, *Elite Oral History Discourse: A Study of Cooperation and Coherence* (Tuscaloosa, 1990), pp. 114–116, for discussion on perceived need to create an unobtrusive interview setting.

10. A wide range of projects exist on a nationwide basis, including the National Park Service (Jimmy Carter National Historic Site), where the former president and first lady were interviewed; the National Press Club's "Women in Journalism" project; the Holocaust Museum's work in recording the memories of survivors (to be edited for exhibits); the Living History Program at Duke University, where post-WWII American leaders are interviewed; the Minnesota State Historical Society's effort to record farm practices, environmental issues, and the resort industry; the University of Nevada/Reno's projects to record Indian and ranch life; Winthrop College's project (South Carolina) to create a unique historical resource in women's history; the Mendocino County (California) Museum to record visual portraits of citizens and to document collections; the city of Grand Rapids, Michigan, to record local personalities; and the Maryland National Capital Parks and Planning Commission's history projects to document African-American settlement.

11. Memo from Stephen White to Arthur Singer, October 6, 1985; *1982 Annual Report*, Alfred P. Sloan Foundation (New York, 1982), p. 42.

12. *1984 Annual Report*, Alfred P. Sloan Foundation (New York, 1984), p. 52–4.

13. 1983 and 1984 *Annual Reports*, Sloan Foundation.

14. Letter from Nathan Reingold to Robert McC. Adams, June 20, 1985.

15. Memo from Roger Kennedy to Robert McC. Adams, July 3, 1985.

16. Memo from William W. Moss to Secretary Adams, June 21, 1985.

17. Memo from DeVorkin to Pachter, Moss, Reingold, and Fleckner, November 18, 1985, regarding the Sloan Video History meeting of 11/15/85.

18. Memo from David DeVorkin, January 3, 1986, to Video History Committee Participants, regarding Progress Report and summary of Decem-

ber 9 meeting; memo from David DeVorkin to Distribution, March 3, 1986, regarding the Sloan Meeting of February 28, 1986; memo from David DeVorkin to Distribution, March 27, 1986, regarding Sloan planning.

19. Proposal to the Alfred P. Sloan Foundation to fund the Smithsonian Videohistory Program, May/June 1986.
20. David DeVorkin, Sloan Videohistory Report #1, October 19, 1986.
21. Devorkin, Sloan Videohistory Project Report #1, October 10, 1986.
22. See Charles I. Weiner, "Oral History of Science: A Mushrooming Cloud," *Journal of American History*, 75(September 1988), pp. 548–59, for a further discussion of ideas regarding a historian recording contemporary events, in anticipation of their value for the future.
23. DeVorkin, October 19, 1986.
24. Memo from William Moss to David DeVorkin, December 4, 1986; an exception was when David Allison interviewed J. Presper Eckert about the ENIAC; the interview was shot experimentally for both general documentation as well as for the National Museum of American History exhibit, The Information Age.
25. Selma Thomas, Report on Twentieth Century Small Arms videohistory project, interview with Eugene M. Stoner. May 1988.
26. Memo from Moss to DeVorkin, Reingold, Pachter, and Henson, February 13, 1987.
27. See Appendix 1 for project summaries of Smithsonian Videohistory Program.

2
PRODUCING VISUAL EVIDENCE AND VIDEOHISTORY: CONSIDERATIONS IN APPLYING A NEW HISTORICAL RESEARCH TOOL

Scientists, anthropologists, and psychologists, among other scholars, have incorporated and analyzed visual images as evidence in research for many years. Initially, photographs supplied a rich record of visual data, but as film and video became available, scholars began using moving images. In the late nineteenth century, the father of American anthropology, Franz Boaz, used film to document traditional lifeways, and by the 1930s and 1940s anthropologists Margaret Mead and Gregory Bateson used film for analysis of cultural behavior.[1] Jake Homiak, ethnologist in the Smithsonian's Human Studies Film Archives, noted that these early ethnographers did not have the luxury of a portable and relatively cheap documentation tool like the video camera. Therefore, when they used film they chose to document dance, ritual, gestures, and movement, rather than "talking heads."[2] Since that time anthropologists, in the study of visual anthropology, have expanded the use of film and video as a research tool to assist their investigation.[3]

Folklorists, too, used film and video to understand their subjects. For example, the Smithsonian's Office of Folklife looked at the use of moving images as an adjunct to printed words, photographs, and verbal descriptions of culture to gain a visual, expressive mode of communication that, when added to existing scholarship, completed its full dimension. "What we witness is performance . . . with sound and video, and through it we hope to capture the complete flow of events, character, speech patterns, moods, and personalities. . . . "[4]

Historical archaeologists and historians of material culture have also incorporated visual evidence, such as structures, places, and objects, as an integral part of their research. Scholars such as James Deetz and Thomas Schlereth have addressed skills needed for using and analyzing visual evidence; Deetz analyzed architecture and artifacts for use in interpretive programs, while Schlereth claimed that objects provide historians with "numerous and valuable insights into

19

the past. To neglect such data in any modern historical inquiry is to overlook a significant body of research evidence."[5] They, and others like them, are adept at asking questions about objects and places, and at evaluating and interpreting such items as evidence. Other humanities scholars, such as Barbara Carson and Gary Carson, addressed the issue in the realm of social history when they noted that "historians who have a clear knowledge of material life have a clear advantage over those who have not."[6] These are among historians who have addressed the value of visual evidence and the role it plays in collecting and interpreting historical data.

Traditional academic historians, however, have focused on written documentation as the primary means of gathering evidence. While some include oral narrative to "fill in the gap" of the written record, and use still photographs as illustration and evidence, historians have been largely text-bound for both documenting and interpreting the past. Carson and Carson encourage academics to face a general re-education in historical thinking, and to engage with colleagues of material culture to expand the narration of history. Others, like John O'Connor and Hayden White, advocated the incorporation of film and video, such as movies and docudramas, in research, education, and scholarly analysis. The latter has been a fairly recent phenomenon,[7] although some oral historians, as noted in Chapter One, began to think about using film as another means of documentation as early as the 1960s.

The Smithsonian Videohistory Program built on these previous efforts by determining how to capture historically useful visual information using a video camera. We assumed that videohistory was a potential extension of both the oral history and photographic traditions, where the medium could record conversation and motion over time as well as visual documentation of a person or place. Studies by scholars in other fields suggested how this could be done. We applied material culture study techniques in asking questions about objects, incorporated the knowledge of oral historians on eliciting personal reflections, and used the ability of archaeologists to record and evaluate structures and environments. We also pursued rigorous historical inquiry in subject matter when developing videohistory projects. We had to go, however, beyond both academic and material culture historians, to develop a working knowledge of video technology and how it affected our results.

We wondered what video might or might not add to the historical record. We asked: Why use video? What will it add to the record? Will it complement, not duplicate, other sources? Will it capture unique visual information? Is video necessary for that purpose? Will video augment, not distort, the record?

We were not convinced that videohistory methodology would, or could, be effective for many research situations, and tried to find where video worked the best . . . and where it did not. We found that video is generally most useful when recording the interaction of people with objects, places, or other people, when capturing personality and "body language," and when exploring a process or documenting the function of artifacts. Video is least useful in historical research when people talk about abstract concepts, which is better left for audio or written text.

Our results, based on all twenty-two projects, suggest that videohistory is a potentially significant historical resource. Our experiment refined the emerging notions of videohistory process and how it might be evaluated and incorporated as visual evidence.[8] By placing these projects in the context of established SI research efforts, we were able to decide if the projects really gathered new information and if videohistory is a valid source beyond simply supplementing traditional written or oral historical evidence.

Could videohistory, if grounded in research and analytical inquiry, add anything new to the record? Ronald G. Walters, professor of history at Johns Hopkins University, reviewed samples of the SVP collection and answered affirmatively when assessing the long-term value of such visual evidence. He stressed that historians, in general, should look seriously at priorities assigned to different kinds of evidence, and that although traditional sources may have shaped their understanding of visual evidence, there may come a time when visual evidence gives validity and meaning to other evidence. "What," he questioned, "validates what?"[9]

Projects that focused on process, environment, work style (action, mannerisms, body language), and interaction added a level of information that is unobtainable by written or oral documentation. Videohistory, at its best, is a cooperative effort that requires good management, professional technical support, and historians willing to explore the complexities, nuances, and infinite possibilities of visual documentation.

We also discovered, however, that video technology, when used in research, can influence or drive the intellectual process if not undertaken with careful planning and assessment. Technical choices, such as recording format and the number of cameras, as well as whether we hired directors and crews, affected the quality of the final visual document. Those often expensive choices taught us that videohistory must be used prudently and that the cost of conducting videohistory must be weighed against anticipated benefits.

APPLIED METHODOLOGY:
THE CONDUCT OF VIDEOHISTORY

The conduct of videohistory involved a new methodology for historians. That methodology combines thorough knowledge of content with careful planning, use, and understanding of the medium. Issues such as how to use video to accommodate interview styles, how to incorporate artifacts and sites most effectively, and how arrangement of interview setting determines the quality of information recorded, all play significant roles in the conduct of videohistory. We wrestled with these issues in a variety of circumstances, which ultimately provided the experience necessary to reach sound conclusions and to frame reliable advice. We experimented with levels of preparation, styles of interviewing, diverse locations, uses of artifacts, and group versus one-on-one interviewing to judge about when, where, how, and why to use video in historical research.

Planning, which involved the historian, a staff member, and usually a professional video director, required more time, more strategy, more thought, and more logistical arrangements than most historians anticipated. As noted in Chapter One, SVP staff handled logistical and administrative arrangements, while the historian developed content and ideas for accomplishing the project, and consulted with the producer regarding technical requirements.

We suggested that historians conduct audiotaped oral history interviews when possible because it provided a base of abstract information upon which they could identify the more visual and concrete aspects of a project. Those who had conducted interviews well in advance of their videohistory projects, as well as those who interviewed people fairly close to the video shoot, usually had a better sense of what needed visual documentation, particularly if they were interested in a specific person's work, or in a group of people. Historians more interested in documenting a site or process—and less interested in the biographies of people who performed the work— were less likely to prerecord with audio. Indeed, the lack of it did not adversely affect the final product of these projects.

Levels of preparation and planning were distinctly different among all historians, and this carried into production. Several worked from a planned agenda, almost a "storyboard," while others preferred a more casual mode of interviewing. The first approach told the historian and director when to insert visuals, when to change sites, and what topics to cover, but did not list specific questions. With the more relaxed approach the historian knew what he or she wanted to cover, and let the "chemistry" of the session determine what would be captured on tape. Both approaches often resulted in a nonscripted, almost free flowing style of taping that seems characteristic of videohistory, i.e., taping the action or interview without the frequent interruptions more common in commercial television.

The amount of the historian's preparation, however, did influence the quality of visual documentation recorded. Planning for the visual element of a project was crucial, and we were not always successful in helping the historian identify what aspect of the story would benefit from the moving camera. Those that knew what they wanted spent less time on extraneous information and more carefully addressed the visual aspects. This was obvious on the tape produced. On the other hand, those who had not prepared so thoroughly nor were willing to explore such options with us, spent a good deal of time recording nonvisual information—data which should have been confined to audio recording.

The interviewer's role was to elicit, encourage, and facilitate the taping, not to play a lead or even an equal role. Thus, the approach to an interview was intertwined with two significant stylistic questions: 1) how regularly should a historian be included in the shots? and 2) how intrusive should he or she should be during the interview? In several early projects, such as the Manhattan Project and SI Paleontology, historians were recorded as regularly as the participants. Images of them asking questions, nodding in response, or handing an object to the participant added little to the record and took away time from the most important aspect of the session, the interview subject(s). As a consequence, historians avoided this as much as possible; they did, however, introduce the session, or include themselves in an "establishing shot" that showed them for the record. Most were quite happy with this arrangement since it both prevented them from becoming a disembodied voice during the interview and from taking center stage as moderator.

The inquisitiveness, or intrusiveness, of a historian during an interview was also an issue. While some of this depended on the person's interviewing style, much rested on how comfortable he or she was with the medium. Some took to this mode of inquiry quite naturally while others were more intimidated. Extenuating factors, too, can influence the outcome. One historian arrived for an early afternoon interview in California after an all night cross-country flight and had had little time to spend with the participants ahead of time. He had hoped to review artifacts with the interview group before the session. No amount of advance planning could have resolved the late scheduling conflicts between the historian, participants, and use of a significant location. In addition, the air conditioning was broken at the taping site, and, as a result, the company decided to close early that afternoon. Although usually comfortable with the medium—and certainly flexible enough to deal with the situation—the hot and tired curator was none too ready to ask probing questions. When evaluating this project, SVP committee member William Moss inquired whether there might be something in the video medium that inhibits activist inquiry. Although we were unable to answer his question specifically, we found that, more often than not, circumstances, personality, and preparedness were more significant in the creation and result of videohistories than the medium itself.

Others experienced situations that resulted in less active inquiry. Two historians who conducted a group interview about activities at a major research corporation remained relatively quiet because of constraints placed on them during production. The corporation hosted the videohistory shoot at its headquarters, and the senior vice president was instrumental in gathering participants for the daylong event. The site was selected because of familiarity; both the vice president and historians wanted the elderly participants to feel as comfortable as possible, rather than place them in a more neutral studio setting. The vice president was also a participant and co-moderator. The group arrangement was unfortunate, as it placed the senior executive at the apex of the group. His presence, including his enthusiastic questioning and prodding, virtually dominated the session. He did not, however, ask that participants respond to suit the corporation. The resulting video captured spontaneous answers by, and interaction between, participants who had worked at the corporation during the same period of time and had headed various research departments.

Remaining quiet can have benefits, too. A historian who was usually a probing interviewer remained virtually silent during a taping session with wives of scientists. The women, who were thrilled to be given the opportunity to talk about their experiences, questioned each other, laughed among themselves, and conversed for over two hours. If the historian had been more assertive in questioning, the interview might not have been as spontaneous nor would it have captured the enthusiastic manner of the interchange between the participants.

Another historian, when interviewing a small arms weapons designer, actively helped the interviewee with names, dates, places, and other specific pieces of information. These suggestions, noted the reviewer for the project, "didn't work very well; they were too intrusive on the man and his thoughts."[10] In this case, the interviewer's active interview style seemed, to the reviewer, to interrupt the natural flow of the interview. To the historian, such additions were natural. He knew the interviewee very well, knew his personal history, and was willing to assist when needed. He may have know too much about him to let the story flow naturally.

Most historians, however, balanced their questioning style enough to prompt the subjects without dominating the interview, or without vanishing into the medium. In general, the most effective style occurred when the historian took an assertive role by guiding the interview with direct, concrete, object-oriented questions, without becoming overly intrusive.

Historians who had thought about the type of interview arrangement, such as one-on-one interviews or groups, and the use of objects and environments in a session, were usually more satisfied with the results of their videohistory sessions than those who left these things to the last minute. Also, those who worked closely with their producers seemed to be the most satisfied with the results. In such cases, producers more clearly understood the implications of choices for historical documentation.

One-on-one interviews or groups of two provided some of the SVP's most effective footage. Usually shot with a single camera, these sessions were often more flexible and less intimidating than complicated multicamera group shoots. Participants were generally more relaxed and ready to demonstrate the use of objects or to discuss the importance of a site. Most surprising was the impact video had on certain participants. Historians had expected that a participant who was introspective and shy would behave the same while on camera. This, however, was not always the case. Scientists who rarely talked about themselves or their professions opened up when the camera rolled tape. A reticent astronomer opened up and "performed" on camera. The project's producer noted in his report that "the personality revealed may be more open than in real life. What we may be losing in terms of the strictest verisimilitude, however, may be only in exchange for greater insight into the scientist's work and attitudes. Rather than inhibit, video seemed to serve as an incentive for many people to open up."[11]

The success of a one-on-one interview depended, again, on the

The video camera encouraged Charles Worley, left, to talk about his work as an astronomer at the U.S. Naval Observatory for the Classical Observation Techniques Videohistory Project. David DeVorkin, right. [SVP Photo, Terri A. Schorzman].

amount of preparation the historian gave to the project, particularly to avoid recording a "talking head" interview. The most successful historians appreciated that one-on-one video was most effective when participants interacted with environments or objects, which usually yielded visually important material. In these instances, the flexibility of a single camera to zoom in on artifacts allowed the historian to capture design detail and functional elements of an object, as well as to follow a person around a site while he or she spoke at length about the place and use of objects.

We had only moderate success with group interviews consisting of three or more people. We had assumed that video might reveal social dynamics of groups or stimulate a collective memory within groups who had a common experience. More often, however, group discussions, which usually took place in a studio or other neutral site, pro-

A single camera followed G. Arthur Cooper through collections areas and offices spaces, which allowed him to demonstrate unique and intricate features of his photographic "apparatus" for the Smithsonian Paleontology Videohistory Project. [SVP photo, Terri A. Schorzman].

voked interesting dialogue and reactions, but failed to produce the same degree of richness of information and insight as sessions conducted one-on-one in laboratories, collections areas, or office spaces. Group sessions usually documented the dominance of a leading personality, even many years after the events under study, and they occasionally captured exciting interaction and conversation among members. Generally, however, historians found that group interviews were less useful to the goals of their research than the more concentrated, controlled, and directed one-on-one or one-on-two interviews.

For example, one historian set the tone of his group session by asking questions specifically to each participant, as though he was conducting individual audiotaped oral history interviews. This, in turn, caused members of the group to respond only when specifically questioned. This structure, according to a reviewer, "invited each participant to fill in what he knew best in an overarching story line, rather than to address common issues and experiences from unique and possibly conflicting perspectives."[12]

Sometimes an interviewer was unable to obtain effective interaction and stimulating conversation, particularly when participants failed to challenge each other or were too polite to interrupt. During one project a participant confided that group members had agreed to talk about the "good stuff," rather than get involved in past disagreements, arguments, or failed efforts. He said that only in a one-on-one interview would the historian get to the heart of the story.[13]

Groups were more successful in presenting new information when the participants were informal and well acquainted, or when they had never been interviewed before. Scientists often resisted deviating from a story they may have told many times, even when presented with new information or when questioned by another group participant. On the other hand, participants who had never told their story in a public setting more quickly questioned or challenged other group members. This was true when California software designers convened or when wives of Manhattan Project scientists gathered to tell their story about life in a controlled scientific community.

Historians conducted group interviews when they wanted to obtain different perspectives. A variety of reflections about a corporation's research and culture in the 1950s and 1960s emerged, for example, from interviewing former department heads. This project's two historians hoped that group members would stimulate each others' memories by questioning recollections and challenging assertions. Unfortunately, it did not happen as much as anticipated due to a dominant participant and the historians' non-intrusive interviewing style. Yet, the group of California software designers challenged and prompted each other, which was partly due to both the historian's informal style and the circumstances of the session. In addition, no one member of the group dominated nor was viewed as a mentor. The group interview, in this situation, recorded interactions among colleagues.[14]

Another curator captured how teams of student engineers worked together during robotics design and development for their current class assignment. A single camera recorded daily activities of teams as they worked on their robots in several settings, from a laboratory and machine shop, to classroom discussion.[15]

Groups were less important when the historian simply chose the format as a convenient way to bring a number of participants together for a one-time interview, particularly when they had never shared an experience, event, or working relationship.[16] Groups conducted under this guise not only lacked interaction and spontaneity, but were similar to oral history one-on-one interviews, where an interviewee responds to a direct question. The other participants remain quiet until asked a question. Groups were more effective when they shared an experience, did not contain a mentor, or when they discussed objects.

The use of artifacts and environments during a video session was crucial in the creation of effective documentation. We encouraged historians to "think visually," by deciding how they might use objects, places, and materials to best illustrate or explain the results of scientific thought and technological invention. Those who asked concrete, focused, questions were the most successful at obtaining visual information. Our video projects that focused on scientific theory or used abstract, open-ended questions offered little for visual documenta-

Undergraduates at the University of Maryland worked on their walking robot, designed for national competition, during the Robotics Videohistory Project. The single camera followed them from lab to machine shop to classroom discussion. [SVP photo, Terri A. Schorzman].

tion. Such interviews, particularly if recorded as singular talking heads, were better left for audio. Pamela Henson, historian for the Smithsonian Institution Archives, conducted audiotaped interviews "which tend to be on a fairly abstract and general level. Audio interviews do not capture how scientists go about doing their work or the role objects or technical processes play." To Henson, "video added a qualitatively different type of information . . . visual information."[17]

DeVorkin also recognized the importance of this once he conducted his first video interview. "Unlike oral histories, there is far too much going on to have to rely continually on notes or complex conceptual questioning. Keep it simple and direct. Be clear with subjects about what you hope to gain through interviews."[18]

When historians used the techniques successfully, they usually found the interaction between people, objects, and environments to be the most memorable aspect of their video projects. For example, National Air and Space Museum (NASM) historians were pleased that a specially made photoreconnaissance camera stimulated memories of team members about a project that took place over forty years ago; the objects provided a visual focus that encouraged discussion. One of the scholars, Joseph Tatarewicz, noted that the footage of the instrument with designers "disassembling and commenting on the design and history of operations of the instruments was gold. It will be the major lasting contribution of this shoot."[19] The project's reviewer project also found that the description of artifacts was the most interesting aspect of the interview. "One of the great strengths of the video was that it gave us a chance to see an artifact and have it explained by the people who designed, built, and used it . . . the video gave us a rare look inside some pretty sophisticated and otherwise inaccessible equipment . . . the artifactual analysis was a valuable contribution."[20]

Jon Eklund, curator at National Museum of American History (NMAH), noted that by documenting the interaction between software designers as they operated a computer, he captured the enthusiasm and innovation that excited the industry and made it successful. These issues were portrayed effectively through the inventor's use of an artifact. To Eklund, video seemed to capture the inherent and intimate connection between artifact, purpose, and creator.[21]

Carlene Stephens, also a curator at NMAH, agreed that documenting artifacts was significant in video recording. She found that the videotaped record of her series at the Waltham Clock Company added detailed information to the museum's holdings regarding the clock company, but that "the objects in our collection are static. Seeing the machines as part of the manufacturing process in the hands of skilled workers gives them much-needed context."[22]

Incorporating artifacts in an interview also added new data regarding small arms design and manufacture. With the artifacts at hand, the interviewer, Edward Ezell (NMAH), encouraged the weapons designers to illustrate structural features, to manipulate and rotate weapons for better viewing, and to take apart weapons to demonstrate how quickly the task could be done in the field. The designers constantly touched the weapons, and seemed to those present as though they read the objects with their fingertips. The camera captured their every movement. When a scholar evaluated a tape from the first session, he noted that the artifact, an AR-10 gun, motivated the designer to talk about the weapon's design. "This technique is

Small arms designers Mikhail T. Kalashnikov, left, and Eugene M. Stoner, demonstrated aspects of their weapons designs during the Twentieth Century Small Arms Videohistory Project. The men were brought together for the first time during the session in May 1990. [SVP photo, Alexander B. Magoun].

superior to an oral history interview which moves through the individual's life chronologically. Why is that? The advantage lies in seeing the inventor in the workshop, holding his invention, and talking about what the real issues were in trying to make it work right."[23]

David Allison, NMAH, used artifacts during his interview with J. Presper Eckert, codesigner of the ENIAC digital computer. He found that it allowed Eckert to explain the design and operation of the machine as he referred to it in detailed description. The result was an inventor's explanation of a thought process applied to a technical process. Allison believed that Eckert's philosophy was "realized in the artifact."[24] The project reviewer noted that Eckert provided information that was not available in any other form for the device (accumulator digit storage units); "it is a recording of a man, proud of his technical accomplishment, describing it in terms that make that technical triumph clear. . . . "[25]

In addition to artifacts, video also captured working environments

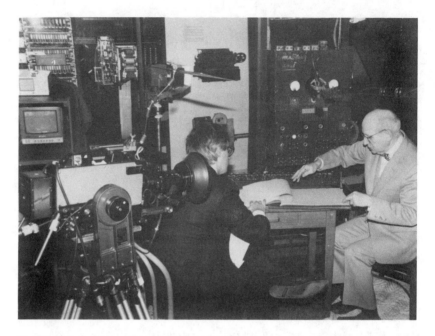

David Allison, left, and J. Presper Eckert, right, talked about Eckert's role in designing the ENIAC computer. Various parts of the computer were available for discussion. [SVP photo, Phillip R. Seitz].

and manufacturing processes. P. Henson interviewed paleontologists in their working environment with their specimens and tools. She found that the artifacts stimulated discussion and influenced the portrayal of objects.

Capturing environment and process was also important to Ray Kondratas, NMAH, during his medical history projects on imaging technologies, DNA sequencing, and cell sorting. He traced the development of imaging, sequencing, and sorting machines from inception to use by interviewing scientists and technicians in the lab and the manufacturing site, as well as medical doctors who used the machines daily. The environment, to Kondratas, was an effective way to document the history of medical technologies. Another historian from NMAH, Peter Leibhold, also used location to significant documentary ends during the New United Motors and Manufacturing, Inc. (NUMMI) videohistory project. He studied the affects of Japanese manufacturing techniques on an old-line U.S. plant, and liked the way video documented people within the work place, and how

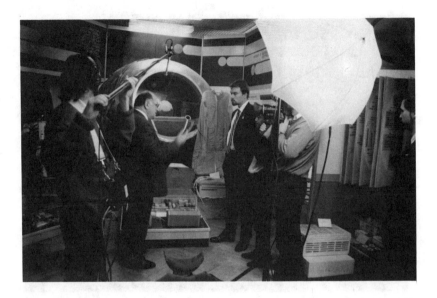

Artifacts played a key role during Cathleen Lewis's videohistory project on Soviet space medicine. Scientist Abraham Genin led a tour through the museum of the Institute for Biomedical Problems in Moscow, Russia. [SVP photo, Phillip R. Seitz].

it recorded both teamwork as well as the adjustments to changing factory life.[26]

Other historians used video to document both fading and emerging industries. They looked at working conditions, process, environment, and people. For example, Carlene Stephens and William Worthington looked at two fading industries that still employ nineteenth century manufacturing techniques. Both captured people working within those environments, including the way they operated equipment and performed duties on site. In an almost *cinema verité* taping mode, Stephens and Worthington recorded step-by-step processes, the precise methods of making a mechanical clock in one case, and the large scale activities of a slate quarrying operation in the other.

Steven Lubar, however, was more interested in documenting emerging technologies than in producing a retrospective. He interviewed engineers about their work at university and corporate sites, where he explored robotics design, manufacture, and application. He also looked at how concepts, through teamwork and troubleshooting, became reality.[27]

Scientist and inventor Robert Ledley was recorded in several environments for the Medical Imaging Videohistory Project, including his office, laboratory, and the National Museum of American History, where his prototype imaging machine was on display. [SVP photo, Phillip R. Seitz].

APPLICATION OF VISUAL EVIDENCE

SVP projects helped establish methodology for creating videohistory documents. Historians experimented with artifacts, sites, and groups, in an attempt to determine how visual information could be gathered most effectively. They were also interested in the application and use of their materials, as they might be with any archival source. Since we supported only production, not post-production editing, historians had to obtain additional funding to develop applications. By the fourth year of the program, the demand for use of the videohistory collection was significant. Scholars, educators, media producers, and others requested copies of the videotapes and transcripts for various purposes.

The most immediate application of videohistory material was for use in ongoing research—to record and collect new information. The majority of the historians undertook their projects primarily for the collection of primary source material and never intended it to be used otherwise. This seemed problematic for one historian, who won-

Peter Leibhold, center, recorded the working environment at NUMMI and interviewed many employees about their jobs; a number of them demonstrated various aspects of the manufacturing process. [SVP photo, Phillip R. Seitz].

dered how to use video documentation as a narrative source. He was concerned that his extensive work in videohistory would count for little when it came time for performance evaluation. How would he, he asked, use the visual evidence he collected in written form? This was an important issue because his work performance was evaluated by written scholarship; his video work was highly visual and although elements could be referred to or quoted in a narrative, the essential component was visual information. Full transcription did not capture the unique aspects of teamwork and action.

Projects such as the Naval Research Laboratory, Smithsonian Paleontology, Mariner 2 launch to Venus, and the series on the RAND Corporation were among projects that were shot almost exclusively as primary research, although some have been used, or proposed to be used, for additional purposes. Professors of paleontology expressed interest in the SI video footage for their classrooms, and the RAND Corporation incorporated footage in a program celebrating their forty-fifth anniversary. Both David DeVorkin and Stanley Goldberg integrated their videohistory inquiry about the Naval Re-

William Worthington, crouched at left, and director Brien Williams, recorded both manufacturing processes and the working environment at Vermont Structural Slate, in Fair Haven, Vermont. [SVP photo, Phillip R. Seitz].

search Lab and the Manhattan Project into forthcoming books. They discuss this process in their essays in Chapter Four.

Other historians anticipated that footage might be incorporated in a post-produced program. Ray Kondratas, for example, undertook a medical history project to document the cell sorter in conjunction with a corporation that wanted to use the footage for an in-house educational program. Other historians, such as Ed Ezell, developed a video series with the possibility that footage might eventually be incorporated into a public program or documentary.

Incorporating footage in an exhibit held great interest for several historians. Goldberg hoped to incorporate his footage in a proposed exhibit on the Manhattan Project, Stephens planned to use segments of Waltham Clock Company tape in an exhibit on clockmaking, and Leibhold shot specific footage at NUMMI for inclusion in an exhibit on Japanese management and American manufacturing. Allison's interview with Eckert was incorporated as an interactive video for the NMAH exhibit "The Information Age." Taping for both archival documentation and exhibit or program footage posed a challenge to

Pamela Henson, center, combined historical perspectives with contemporary scientific practice to document tropical biology at the Smithsonian Tropical Research Institute in Panama for her Conservation of Endangered Species Videohistory Project. Here, Norman Duke talks about his work with mangrove swamps. [SVP photo, Pablo Jusem].

historians and producers. Historians often balanced both interests by focusing on interview content, while letting the director and crew capture extra footage that covered reaction shots, special closeups (known as B-roll to a crew), and other images that would ease the editing process. Occasionally, elements of projects were taped twice, once with vocal explanation and once without it. This, presumably, would provide more flexibility for editing the material in the future.[28]

Finally, videohistory was excellent for documenting industrial processes for museum collections. Worthington's work at the Vermont Slate quarry was a perfect example. He recorded a nineteenth century cableway system of hauling slate from a quarry—an operation not in use for at least a decade. The retired cableway operator returned to demonstrate the system before it was destroyed. Due to its size, Worthington would not have been able to include the system in the museum collection; video documentation was the next best thing. Stephens faced a similar need. She wanted to find out how equipment, which the museum had obtained the 1950s, actually worked.

With video, she was able to document late nineteenth and early twentieth century machinery in operation, with the operators, before it was destroyed at the factory.

ISSUES IN VIDEOHISTORY

During all phases of production, we addressed issues that were important in selecting, creating, defining, evaluating, and using video documentation. Many of these concerns were addressed while developing the Smithsonian Videohistory Program as well as during preparation for individual projects. Most concerns dealt with logistical, theoretical, and ethical issues.

Logistics for videohistory were vastly more complex than preparing for audiotaped interviews and we were often overwhelmed by the amount of preparation and post-production followup needed to ensure a successful shoot. Because of dealing with a large federal bureaucracy, it often took two to three months to pull together all elements of a single video shoot, including the preparation of technical service contracts, paperwork for travel, and other logistics. Although time-consuming, such paperwork was necessary before taping could proceed. Well-conceived contracts were essential for working with professional crews since it let all parties involved understand project requirements. The SI contracts office suggested language; we then created a standard contract that we modified for each project, depending on equipment and service needed. As we became more experienced—and understood the various technical elements that comprised a shoot—our contracts became more specific. A detailed contract assured us that we would receive what we asked for.

Historians prepared for the content of the shoot, contacted participants, scheduled taping dates, and met with the producer/director and staff regarding production requirements. Pamela Henson, for example, found that she spent three weeks planning the intellectual component of her video sessions—in addition to recording audiotaped interviews well in advance of videotaping. "My major observation is that it is vastly more work to set up a video session than to plunk down an audio tape recorder. The researcher must weigh if the session warrants all that work and fuss and bother."[29]

Historians, particularly those who had conducted many oral history interviews, were occasionally disturbed by the lack of privacy during a videohistory shoot. They were used to one-on-one audiotaped interviews in the privacy of a home or office, where a tape recorder was the only equipment present. Some historians were concerned about what impact the presence of crew and equipment would have on an interview. One in particular did not like the extravagant production of his multicamera group interview and said that he preferred the quiet simplicity of his audio tape recorder. It was his last videotaping session. Another interviewer was sufficiently concerned about the intrusiveness of a crew and production assistant that he obtained a camera operator the participant knew well and asked that no one else be present during taping. The interview, even though a "talking head," was very good. It lacked, however, the supporting data of a tape log, production notes and set design, good photographs, and a signed release form—tasks usually performed by a production assistant.

Most of the participating historians, however, found that while interviewees were slightly uneasy at the start of a recording session, they soon relaxed and paid little attention to production. This was true across the board, particularly for those who demonstrated their work in a familiar environment or conducted a tour of a site or process; they became so engaged in the discussion that they paid little attention to the camera and crew. Again, had these participants been placed as a "talking head" before a camera and asked to reflect on personal issues, they would have probably been more uncomfortable and not as forthcoming.[30]

Another issue in recording videohistory was deciding who or what to record, since we assumed that the selection would help tell the future about life in the twentieth century. Although SI historians selected topics based on their ongoing research, the advisory committee emphasized that material should be collected for indefinite future historical interest rather than collecting for specific and immediate purposes. They preferred that historians "cast a wide intellectual net." A few historians thought this approach was too extravagant; that producing an exhaustive archival collection of video footage without a specific use in mind was not worthwhile. Others argued that highly focused productions would miss potentially valuable historical information, which would, in turn, limit its use as a true archival resource. Ezell encouraged the use of video as a historiographical and archival tool by noting that much valuable documentation required for a full picture of his field of study, military technology, was either never collected or never existed. To him, the availability of video technology made it possible to capture both the visual and oral aspects of people and the technology they developed.[31]

Other historians, while they agreed with the broad emphasis on archival collecting, were troubled by recording groups of people or situations that seemed artificial or staged, particularly when interviewees gathered at a relevant historic site. For example, Goldberg returned two aging physicists to the Trinity site, where the first atomic bomb was exploded, to get their reaction to the place after almost fifty years. He hoped that they would document, or "read" the site for him. The physicists reminisced about what happened there as they walked around a familiar, yet distant, place with a friendly collegiality. The effect was "staged" yet impressions were spontaneous. In other situations, the taping process was a reenactment or recreation, but seemed no less real in its documentary power. For example, DeVorkin encouraged reenactment when he recorded the observing techniques of an astronomer at the U.S. Naval Observatory. The astronomer actually observed stars at night when light levels were not sufficient for taping, so the crew recorded during the day. To remedy the situation, the astronomer proceeded through his regular observing routine, as though it were night. The reenactment, while not an authentic observing session, still documented the unique practice of the astronomer. The video and supporting documentation, however, stated very clearly what had taken place and why. In every case we remained true to the source by supplying supporting documentation so as not to deceive researchers about creation of the footage.

Ultimately, historians hoped for open responses and spontaneous reactions to their questions. As a result, we learned how to balance the most delicate aspects of recording, neither encouraging a performance nor inhibiting a forthcoming personality. This was not always easy, nor always successful. If the historian remained sensitive to the participant, thought carefully about the approach to the project, consulted with the director about the amount of equipment needed to obtain sufficient visual and audio quality, and considered the long-term impact of his or her product, the result was quite effective.

Historians, then, by creating raw material, took an active role in shaping historical data, whether it was narrowly defined or broadly conceived. Their choices of who or what to record, and how to record it, affected and influenced the result.[32] A future researcher must make his or her own impressions of the material and judge it according to his or her acquired knowledge of the subject. The interviewer's intentions—as well as the interviewee's impressions—can become one more tool for evaluating the record.[33] Yet, evaluating the legitimacy of the video record may become increasingly difficult. The easy and undetectable digital manipulation of images may lead to the demise of authentic photographic sources, or in trusting images as evidence.

Although still images are easier to change than video, it can still be done. The potential effect on history, as well as in journalism and the law, may be the creation of new interpretations due to altered images. The line between entertainment—where audience deception is nothing new—and historical accuracy may dissolve.[34]

Legal and ethical issues were also very important to consider. To establish legal guidelines, we met with the SI's Office of General Counsel at the outset of the SVP to develop a set of release forms. SVP forms covered the concepts of gift, copyright, and physical property. The release form asked that participants transfer their share of copyright in the material to the Smithsonian Institution for the "increase and diffusion of knowledge." It also made clear that the physical property of the tape and transcripts belonged to the SI, and stated that the participant placed no restrictions on use of the materials. The materials, they were told, were part of an archival project in which tapes would be deposited for use by researchers for educational purposes. Participants were given the option of signing restricted forms. See the appendix for examples of forms.

Only one interviewee asked for special restrictions. She did not mind whether or not people viewed the tapes and transcripts in unedited form. What concerned her, however, was the ability for others to edit video in such a way that may ultimately distort her story or the way she chose to represent herself. She gave a very personal interview and was concerned with the way the material might be used, since there is a great demand for such videotaped interviews. The camera, unlike an audio recording, captured every gesture, facial expression, or nuance, which left this interviewee feeling uncomfortable with how much she exposed. Thus, she restricted copyright and stated that anyone who wanted to use or quote from the tapes needed her permission to do so.[35]

When conducting videohistory sessions, we were concerned with the "ethical obligations to interviewees who might later be disappointed or embarrassed by what the video camera 'saw,' especially if it had little or no bearing on the factual or substantive data in the voice-recording part of the interview."[36] During a group session an elderly scientist became so tired that he laid his head on the table. The interviewer and group, however, continued the discussion. Should the interviewer have stopped for a break? In at least two other cases elderly participants were shocked to find themselves looking so old and "doddering," (not the image they had of themselves) while another was embarrassed when he couldn't perform a certain task—and pleaded with the crew to turn the camera off, which they did not. He was worried that the image of his fumbling would remain in people's minds, not the otherwise excellent portrayal of his science. To ease his mind, the material was edited from all tapes, including the master. We did not view this as a significant alteration to the historical record, since it added no new information and was simply an embarrassment to the scientist. Indeed, when "the camera never blinks," the historian must be sensitive enough to know when to stop—and the crew must be willing to follow that request.

In another case, the historian unknowingly told a group of interviewees that taping had stopped, when, in fact, the cameras continued to run. The people may not have carried on the same conversation had they known they were being recorded. We were not aware of the error until we viewed the tapes at a later date. This stressed the importance of a historian communicating with both the participants and with the technicians.

The high cost of broadcast quality video production was also a significant issue. We strove for quality at reasonable cost, but believed that the quality of content, not technological wizardry, should drive the production process. We learned that sophisticated production techniques could often overwhelm the production and trivialize or deemphasize the intimacy and content of the session. For example, during a group interview at NASM, the director chose to use an electronic device, known as a "still store," for displaying closeup images on the screen. While we had a much better view of the photograph or map under discussion, we were unable to see the interaction of the participant with the image as he pointed to a particular site on the map or to specific parts of an object shown in a photo. We wanted the technology to support the production effort, not to be so complex as to hinder research endeavors.

SUMMARY

As the Smithsonian experiment shows, historians have begun to incorporate and analyze moving visual images in their research endeavors. They developed techniques in questioning from both oral historians and historians of material culture, and learned to "think visually" when planning videohistory sessions. By combining both traditional research methodology with video technique, technology, and professional assistance, the SI historians not only addressed the use of video as a research tool, but found that visual evidence could be a valuable source in historical research.

Producing videohistory documentation, however, involved complex logistical and content planning, and was quite expensive.

NOTES

1. Margaret Mead, [introduction] "Visual Anthropology in a Discipline of Words," *Principles of Visual Anthropology*, ed. Paul Hockings (The Hague and Paris, 1975).
2. Memo from Jake Homiak to Pamela Henson, March 18, 1991, regarding visual anthropology. Homiak also noted that Boaz used the "photo-elicitation method" that involved the use of photographs by the ethnographer to investigate the meanings of these cultural forms and/or events.
3. John Collier, Jr., and Malcom Collier, *Visual Anthropology: Photography as a Research Method* (New Mexico) revised and expanded 1990, pp. 139–140; Chris Raymond, "Increasing Use of Film by Visually Oriented Anthropologists Stirs Debate Over Ways Scholars Describe Other Cultures," *The Chronicle of Higher Education*, March 27, 1991, p. A5, A8–9; Richard Sorenson, "A Research Film Program in the Study of Changing Man," *Current Anthropology*, 8 (1967), 443–69.
4. Ralph Rinzler and Robert Sayers, "The Meaders Family, North Georgia Potters," *Smithsonian Folklife Studies*, Number 1, 1980.
5. Thomas J. Schlereth, *Material Culture Studies in America*, (Knoxville, 1982), p. xiv; Schlereth, *Cultural History and Material Culture: Everyday Life, Landscapes, Museums* (Ann Arbor, 1990); Wilcomb E. Washburn, "Material Culture and Public History: Maturing Together?" *The Public Historian* 13 (no. 2, Spring 1991), 53–60, for analysis of the latter Schlereth book and others. Also, see, essay by Carlene Stephens in Chapter Four for a good discussion of material culture studies and video.
6. Barbara Carson and Gary Carson, "Social History From Artifacts," article in *Ordinary People and Everyday Life: Perspectives on the New Social History*, ed., James B. Gardner and George Rollie Adams (Nashville, TN 1983), pp. 181–203. Also, Jeffrey Brown, professor of history at New Mexico State University, spoke eloquently about why historians should include video technology in research, education, and public presentation at the jointly-held annual meetings of the National Council on Public History and the Southwest Oral History Association, San Diego, March 1990.
7. See John O'Connor, *Image as Artifact: Historical Analysis of Film and Television* (Malabar, 1990) for extensive discussion about ways of analyzing moving images. He noted that historians must apply the same critical standards to moving image documentation as with any other historical evidence. As an archival source, according to O'Connor, moving image media can "establish or verify historical facts or be indicators of social and cultural values." See also, Hayden White, "AHR Forum: Historiography and Historiophoty," *American Historical Review*, 93 (no. 5, Dec. 1988), pp. 1193–1199. To White, the "historical evidence produced by our

epoch is often as much visual as it is oral and written in nature. Also, the communicative conventions of the human sciences are increasingly as much pictorial as verbal in their predominant modes of representation. . . . be aware that the analysis of visual images requires a manner of 'reading' quite different from that developed for the study of written documents" (p. 1193).

8. "Annual Report: Summary of First Year Activities and Request for Second Year Funding (FY88)," Smithsonian Videohistory Program, September 1987.

9. Ronald G. Walters, remarks made during panel discussion, "A Visual Approach to Documenting the Past," for the jointly held annual meetings of the Organization of American Historians and the Society for History in the Federal Government (held in Washington, D.C.), March 1990. See also, Walters, "Fixing the Image," *The Public Historian*, 13 (Spring 1991), pp. 101–106, for further discussion about visual evidence and analysis of the SVP.

10. Joseph P. Harahan, "Evaluation of videotape '20th Century Small Arms with Eugene Stoner by Edward Ezell'," November 16, 1988.

11. Brien Williams, Production Report, Classical Observing Methods Videohistory Project, 1988.

12. DeVorkin, Review of 25th Anniversary of Mariner 2 Videohistory Project, 1988.

13. Participant comment, Early Computing at the RAND Corporation Videohistory Project, Paul Ceruzzi, interviewer, June 12, 1990.

14. Martin Collins and Joseph Tatarewicz, Research at RAND Videohistory Project, January 27, 1989; Jon Eklund, Mini- & Microcomputers Videohistory Project, July 31, 1987.

15. Steven Lubar, "Robotics Videohistory Project at University of Maryland and Carnegie-Mellon University," March 31, 1989, and September 20–21, 1990. See his article, "Robot Videohistory," in Chapter Four.

16. Stanley Goldberg explores this concept in Chapter Four, in his article, "The Manhattan Project Series."

17. Pamela Henson, "Visual Documentation and Historical Research," paper presented at the joint meeting of the Organization of American Historians (OAH) and the Society for History in the Federal Government (SHFG), March 23, 1990.

18. DeVorkin, Response to SVP questionnaire, Naval Research Laboratory Space Sciences Videohistory Project, August 6, 1987.

19. Joseph Tatarewicz, Response to SVP questionnaire for RAND Videohistory Project (Photoreconnaissance, Sessions 2 and 3), January 1988.

20. Stewart Leslie, Review of RAND Videohistory Project, 1988.

21. Jon Eklund, Mini- & Microcomputers Videohistory Project, July 1987.

22. Carlene Stephens, Report on Waltham Clock Company Videohistory Project, July 19, 1989. She explores the notion further in her article on the project in Chapter Four.

23. Harahan, "Evaluation," 16 November 1988.
24. David Allison, Response on SVP questionnaire, 1988.
25. Michael Williams, "Review of the J. Presper Eckert Interview Video Tapes," February 1989.
26. Ramunus Kondratas, Medical Imaging Videohistory Project, April 6, 1989, July 5, 1989, October 27, 1989; Kondratas, DNA Sequencing Videohistory Project, October 19 and 21, 1988, March 27, 1990; Kondratas, Cell Sorter Videohistory Project, January 30, 1991, April 19, 1991, June 28, 1991; Peter Liebhold, NUMMI Videohistory Project, September 25–26, 1990.
27. William Worthington documented slate quarrying at Vermont Slate Company, October 1989; Carlene Stephens documented clock making at the Waltham Clock Company, June 1989; Steven Lubar documented robotics engineering at the University of Maryland, March 1989, Odetics, Inc. December 1989, and Carnegie-Mellon University, June 1990.
28. Brien Williams's article in Chapter Four explores these issues at length.
29. Pamela Henson, Response to SVP questionnaire, SI Paleontology Videohistory Project, no date.
30. See McMahan, pp. 114–116 for concerns regarding presence of a camera crew. SVP experience showed that participants rarely paid attention to the camera and presence of others once the interview or demonstration began.
31. Terri Schorzman, "Smithsonian Videohistory Program Symposium," *Technology and Culture*, 30, (no. 1, January 1989), 118–22.
32. See discussion on "reflexivity" in Henson and Schorzman, "Videohistory: Focusing on the American Past," *JAH*, p. 622, as a constructed representation of a work place, process, culture, etc.
33. Page, "A Visual Dimension to Oral History," p. 21.
34. Ideas presented by Don E. Tomlinson and Ray Fielding, seminar "The Computer Manipulation and Creation of Video and Audio: Assessing the Downside," at The Annenberg Washington Program, Northwestern University, December 10, 1991.
35. The Oral History Association "Evaluation Guidelines" states that the interviewer should consider the potential for public programming and research use of the interviews and should endeavor to prevent any exploitation of or harm to interviewees. The Guidelines, as well as the "Principles and Standards of the Oral History Association" are good references concerning legal and ethical issues and practice.
36. Charlton, "Videotaped Oral Histories: Problems and Prospects," *American Archivist*, p. 235.

3
TECHNICAL CRITERIA AND
ARCHIVAL REQUIREMENTS

Creating a videohistory document required a combination of talents and skills. The historian provided the intellectual component that gave substance to the work, the program staff coordinated and arranged logistics, while technical support provided the competence to produce a quality recording. Storing and using videohistory documents also required special criteria to both preserve the materials and make them available for use.

TECHNICAL CRITERIA: SVP OVERVIEW

Because of the experimental nature of the SVP, video sessions recorded during the first four years varied in technical quality. This was due to varying expertise of the technical crew and director, lighting arrangements, sound levels, equipment quality and the degree to which it was tuned, setup, number of cameras, environmental conditions, time of day, shooting schedule, technique in handling artifacts, and closeups. Most sessions were shot with professional crews who responded flexibly to less than ideal settings, including tight office and laboratory spaces, rooms that were not air-conditioned, cloistered scientific collections, and fluctuating electricity currents in foreign locations.[1]

Our experience showed that a skilled crew and a flexible director ensured a better quality product. Technical problems arose frequently in the first two years of the SVP, and professional crews knew how to handle them. By the last two years, we better anticipated potential problems and worked with crews and directors to compensate for possible malfunctions. We also found that how well problems were handled was directly proportional to how well trained and experienced a crew and director were.[2]

SELECTING PRODUCERS AND CREWS

When establishing the SVP, we decided that we would not purchase equipment nor become technical experts about the production process. We also wanted to try a variety of broadcast and near-broadcast quality systems, and to use equipment as it changed in quality and capability. Finally, we did not plan to operate the equipment; we assumed that professional crews, with the guidance of a trained director, would ensure that sound and images were the best possible, which was our goal for creating long-lived, usable footage. In most cases, our assumption was correct.

Selection of both directors and crews were random at first. We met with production houses that supplied crews, spoke with freelance crews, and followed through on references from people experienced in video production. The same was true for directors. There are over nine hundred professional film and television producers registered in Washington, D.C., so we were not sure where to begin. Several had either heard about the program and came to us regarding opportunities, were Smithsonian employees, or had worked with the Smithsonian on other projects on a freelance basis—usually for exhibits or documentaries. One historian had already found a director with whom he wanted to work. Another project, shot at a corporation, provided its in-house director and crew. Again, as we gained more experience, we learned how to select the appropriate director for a project, and then let the director choose a crew. We conferred closely with the director regarding technical requirements and quality of technicians. Our selection criteria follows.

Videohistory, as opposed to standard television production, was a new experience for many directors and crews. To communicate effectively with them about the process, we had to become familiar with production requirements and general terminology. When hiring, we looked at samples of work ahead of time, tried to balance cost with quality (extremely expensive, high-end production crews and directors were unnecessary for videohistory), and attempted to determine the relative comfort of crews when shooting the unstructured subjects of unscripted videohistory by talking with them about the process. Brien Williams, director, was challenged while recording videohistory. He noted that crews found it difficult to produce long, uninterrupted segments characteristic of videohistory, rather than short, highly controlled ones that they were used to in commercial television. Commercial crews also tended to be uncomfortable without rehearsal, but the historians avoided rehearsing to increase the likelihood of spontaneous behavior. Williams encouraged a crew to avoid quick, slick, high impact images and tightly organized flowing visuals, while simultaneously making certain that a high percentage of the footage was useful for future broadcast quality production requirements.[3]

Crews who had worked ENG (electronic news gathering) or who had done extensive field work, were often more suited to recording videotaped interviews in non-studio locations since they were flexible, willing and able to shoot "on the run," and generally more adaptable to unusual circumstances. Those that worked regularly in a studio, for commercial television, were usually less able to adapt to our requests. For example, one person—selected to act as both director and camera operator on a shoot in a tropical location—had assured us that he was experienced with potentially difficult field work. He came with recommendations, was supposedly fluent in Spanish, and knew his way around the Central American country. When we repeatedly asked for his final bid, and for what equipment he needed (so we could list it in both the final contract and for customs) he simply could not deliver. He seemed stifled by the procedure and baffled by what equipment was necessary for the job. We began to doubt his ability to shoot in mangrove swamps and hike into the jungle carrying heavy equipment. We subsequently found someone else, whom we later learned was also more familiar with studio work, where everything was controlled by a tight script. Our work, especially in the field, rarely followed studio rules. Another crew, who had done an excellent job when shooting in a museum, where the lights, audio, and camera placement were arranged with care, failed miserably in the field. The cameraman was extremely slow, was unable to move with the action, and seemed paralyzed without a director telling him every shot to take.

We wanted people with exceptional ability in operating a camera. Good camerawork was, for our purposes, more important than broadcast quality systems. Poor camerawork looked bad, no matter how expensive the equipment. We wanted a camera operator who had the ability to follow action, who could listen carefully to an interview, and who could then respond with appropriate camera direction, such as shooting a close-up when a scientist pointed to an object, or to pan the room when the same scientist indicated special features about his lab. The ability to incorporate graphics and artifacts without "chasing" them with the camera, to keep the interviewees in perspective with the artifacts and environment, to zoom and pan with ease, and to capture gestures and expressions with sensitivity, were the traits of an experienced professional.

We sought people who were easy to work with. On several occasions we ended up with crews and directors who were incompatible with the videohistory approach. Early on, we selected them independent of the director; later on, the director chose the crew and we hired them as a package deal. Crews either came with recommendations or had experience working with the director. We were usually happy with selections although we occasionally had trouble. One experienced news crew, while taping on top of a twelve thousand foot mountain in Arizona, simply refused to take directions from our director, and questioned every aspect of our team's decisions. "The crew was proprietary towards production . . . where they felt we were intruding on their territory. There seemed to be little interest on their part to explore and understand or to see things from our perspective . . . they wanted to push ahead deciding unilaterally how things were to be done."[4] In hindsight, for this case and others, the cost of transporting our own crew may have been worth it. This would have been extremely expensive, however, since we would have paid their daily fee for travel time, work time, and down time, as well as all per diem costs. We took our crew only to foreign locations and to places where we could travel inexpensively (by van, for example) and required only an overnight stay.

Once, when we did take our crew, the camera operator flatly refused to tape additional footage because he was too tired. This would have been understood if the information could have been recorded the next day, which was not the case. This was a one-time opportunity tropical location with a participant who was available only at that time. When dealing with people in situations where the work is quick, intense, and timely, we found the best thing to do was to remain flexible. We could never have planned, nor anticipated, such a reaction.

Other crews, however, showed remarkable stamina, good will, and creativity. Thom Wolf, a cameraman with whom we had developed a good working relationship, went to great lengths to make sure that all aspects of a weeklong videohistory session conducted in Moscow, Russia, were perfect. When the sound deck failed while at a scientific site far from our downtown Moscow hotel, he quickly offered to go to the hotel to get his backup unit. Unfortunately, the driver assigned to the SI group had left the site; Wolf braved Moscow's freezing temperatures and hitchhiked to and from the hotel. He returned, equipment in hand, and ready to continue work. This sort of camaraderie and commitment to a project made working with good crews worthwhile.

Directors freed the historians from making the many decisions about camera angle, image quality, and general composition, allowing them to focus on content. We found that a competent video director not only relieved the historian's anxieties, but anticipated technical problems, advised on session design and setup, and ensured quality in a finished product. The best directors were professional and flexible, their previous work was compatible with our goals, and they understood archival and documentary collecting. We looked for a person who understood that producing videohistory was different from producing the evening news, and who showed a willingness to explore the medium for effective and sensitive footage. Such producers usually had an academic background, either in history or the liberal arts, rather than exclusive training in commercial media.

We encouraged a producer/director and historian to work as a team, and let the two decided how active a role the director would take during production. Variations in the styles of both directors and historians provided a wide range of possible combinations. Some directors asked that the historian govern the pace of the proceedings, while others directed the proceedings, called for reevaluation of scenes, and asked to reshoot things that weren't initially satisfactory.

Several historians encouraged the use of an active director to oversee all facets of technical and logistical arrangements while leaving content and intellectual components to the scholar. Others preferred to be partners in the process, which allowed the historian to design, determine, and structure the final product. The director would, in turn, suggest questions, content, and graphics that were viable for video.[5] For example, one curator developed a close working relationship with the producer by questioning many aspects of the video process during taping. This assured her that she was getting good visuals, that her questions to participants were clear, and that she facilitated, not delayed, the videoshoot. The producer found that the historian's help was important. "We were able to dissect each step in the process, deciding what aspects of that step were critical and which point of view was more appropriate for the camera. She had clearly done her homework . . . and that made our work easier."[6] The director was thus able to suggest where closeups were necessary, prompt the curator to address aspects she might have missed, and assure her that the recording process was going according to plan.

Another historian, who taped for both historical information and exhibit footage, worked with a person who directed all technical arrangements as well as assisted with intellectual content. The director asked questions of the participant during the interview and stopped the process regularly so that the cameraman could change positions, get tighter shots, or readjust the placement of the participant. The historian stated that the director was more knowledgeable about visual information and should consequently be active in collecting material most suitable for public education. A scholarly reviewer,

however, found this technique very distracting. "The constant interruption from the director tended to distract the participant and ruin
what was developing into a fine running narrative. I suggest that, in
the future, any attempt to construct exhibit material from the videohistory project be secondary in importance to obtaining historical material."[7] He also addressed the obvious tradeoff in the shoot, in
choosing the technical competency of an assertive producer/director:

> Contents of this tape were expertly photographed and the audio levels
> were always perfect. It was obvious that the production crew was excel-
> lent and this aspect of the situation should be retained if at all possible.
> I think that the mixture of a technically excellent production with a
> director that is neither seen nor heard may well be impossible to achieve
> . . . if the high quality was only managed by having a well qualified and
> concerned media person there to ensure that everything was correct.
> . . . it is reasonable to presume that they will occasionally need to inter-
> rupt the interview to correct some technical aspect of the production
> or even to suggest that a more informative shot be taken. Perhaps a
> happy medium can be found between the two extremes.[8]

We finally chose directors who were technically competent and intellectually engaged in the project, but who were less assertive during
the taping process. These people were quick to make suggestions,
but agreed with the historian that the interview process should remain more relaxed and not conform to a tightly directed format.

The best videohistory projects came from self-directed historians
who knew what they wanted, who were involved with the director
and staff in the planning and design of the project, who were willing
to assert control over the outcome of the project, and who had a
sense of visual appreciation for the subject. This occurred most regularly with historians who dealt with material culture on a daily basis
and who, as a result, knew what to document regarding a site or an
artifact. A shared level of knowledge on the part of both historian
and director created a better informed and more technically correct
product. The better the product, both intellectually and technically,
the more application it would have.[9]

Overall, we worked with flexible and competent directors who appreciated that videohistory was a distinct, experimental area with
needs far different than shooting for a scripted production. We appreciated directors who helped historians understand the visual element of video and the purpose of using its techniques to accomplish
their goals. These people helped develop our experience, understanding, and sophistication with the medium.

SELECTING EQUIPMENT

Video technology and the components needed to make a successful shoot can often overwhelm a person not versed in technical requirements. Concepts such as composition, angles, and framing; techniques such as zoom, pan, and dolly; and terms such as chyron, sync-generator, and time code can cause historians interested in conducting videohistory to throw up their hands in desperation. Historians do not need to know specifics about video technology, but should become familiar with basic terms and requirements to help ensure a quality production. The more aware historians are of video formats and equipment requirements, the better they will communicate with professionals and the more likely they are to get the desired product.[10] The medium, then, is part of the message. We have become so accustomed to seeing excellent images, that poor quality videohistory footage detracts from the program when later used for broadcast or exhibit. So distracting is poor image composition that it almost prevents the viewer from "hearing" what is being said. So, although the "Gettysburg address is as moving on the back of an envelope as it is on parchment,"[11] those are words being conveyed, not visual images.

We learned how to coordinate a shoot without becoming technical experts ourselves. Working with the director and historian, we selected equipment, decided on the number of cameras that should be used, and determined what was needed to accomplish a shoot. We thought about the site, the number of participants, and in some cases, the use of the final product. Since we were financially able to support broadcast quality production, most projects were shot on higher end formats such as open reel 1″ and high-speed 1/2″ Betacam cassette. One-inch tape was used only in a studio, with multiple cameras, for group interviews. While its quality was very good, its lack of portability limited its use for fieldwork. Betacam, on the other hand, provided the same excellent visual quality, yet was portable enough for taping at all on-site situations. The use of high end production formats gave projects more flexibility for a variety of uses, from exhibits to classrooms, and were more stable for duplication and long-term storage.

Some projects, however, were shot on U-Matic 3/4″ tape, a near-broadcast quality format. This usually occurred when the SVP shared the cost of production with a host institution, which provided an in-house crew and equipment. When this happened, crews were usually less experienced than freelance professionals, and had rarely shot unscripted productions. On one occasion the crew lacked the experience to know how to respond to a director's cues, and was not adept at zooming and focusing to capture the unique detail of a particular object. In another case, an inexperienced in-house director failed to stop recording for a break; the result was a continuous four-hour interview with four elderly gentlemen who were too polite to ask to stop. The historian was so engrossed in the discussion that he failed to note the time; the crew waited for a cue from either the director or historian—and got one from neither.

The use of such crews with 3/4″ tape occasionally resulted in flat visual material, where lighting was too harsh or too weak, or where the director and crew overlooked closeups or reaction shots. Such in-kind services, while helping reduce the cost of a shoot, occasionally weakened the technical quality of the final product. In some cases the SVP, as a guest of a hosting institution, was bound by technical limitations, use of facilities, and good will, to undertake a project in a certain way. This can be a reasonable way to complete a project.

We never shot master tape on 1/2″ VHS cassette, but dubbed all research copies to that format. At least one SI historian tried VHS on his own as a "field notebook" after he gained initial experience with the SVP. It gave him a quick way to record an object for the collection when he wanted visual and narrative description to accompany it. He found that shooting in VHS was a good way to record historical data when additional funds were unavailable for higher-end production, particularly if the tape was not intended for broadcast or any other potential use.

When contracting for technical services, we asked for basic equipment such as a camera, audio deck, microphones, tripod, light kit, playback monitor, and cables. We wanted lighting that was not too dramatic, yet not too spotty, requested that a monitor be on site so that the director, and occasionally the historian, could view images being recorded, and asked that microphones be consistent throughout a shoot to achieve better sound balance.

We also specified that time code be generated on site. We failed to ask for this feature for several early projects when we recorded on U-Matic systems, which do not generate code. As a result, time code was applied later at significant extra expense. The television industry developed time code so that video users could refer to or calculate the length of specific parts of a videotaped recording. Time code is helpful for archival videotapes since it enables researchers to find specific passages, cross-referenced on a transcript, and cite them with the same accuracy as pages in a book. Time code places a number on each frame, a unique, time-oriented "address" that indicates

This image, taken from video footage of Pamela Henson's interview with Theodore Reed at the Smithsonian Conservation Research Center in Front Royal, Virginia, shows the time code window at the bottom. [SVP video, Thom Wolf].

elapsed time between two points. It is laid on the master tape as an electronic signal and can be transferred to a "window" for viewing purposes. We chose to record time code in real time, which indicates the total lapsed time of the interview. A code that reads **01:12:33:14** means that the session, at that point, is one hour, twelve minutes, thirty-three seconds, and fourteen frames long.[12] Betacam units generate time code and additional equipment is not needed.

Extra features, such as multiple cameras, editing machines, and communications systems between director and crew were based on specific project requirements. A director or crew is particularly helpful in deciding what options are available and appropriate.[13]

After selecting the format, the production team, i.e., the historian, director, and a SVP representative, decided how many cameras were necessary for a project. We used one camera for a one-on-one or a small group interview (two or three people), for site tours where we needed the mobility of one camera, for objects and process (as noted

The flexibility of one camera allowed Pamela Henson to capture a variety of activities at the Smithsonian Tropical Research Institute in Panama. Nicholas Smythe conducted a tour of his work with pacas at the Tupper Center. [SVP photo, Pablo Jusem].

in Chapter Two), for quick setup and breakdown, and for more intimate settings. Only two crew members (occasionally three) were present during a one-camera shoot, which was generally less intrusive. A single camera was perfect for these specific situations. David DeVorkin, who had experience with both multicamera group shots and one-on-one sessions preferred a single camera. "I am more comfortable, the crew is more personable, the subject is more relaxed and not overburdened with intrusive people and technology, and we do not have the problem of live mixing [editing]."[14]

We also experimented with using one camera for a four-person interview. The arrangement was not particularly successful, for it required the camera operator to pay close attention to who was speaking, and to rely heavily on the producer for explicit direction. One camera did not permit the recording of speaker and reaction shots, or group interaction. The physical arrangement of the group, lined in a row, also lacked the intimacy that could have been recorded had they been seated differently. The participants were also conspicuously uncomfortable. [See Photo in Goldberg article in Chapter Four].

One camera is effective when space is limited, as was the case for Theodore Robinson's interview with Cornelius Coffey, left, for a series on the history of African-Americans in aviation. Only a library's auditorium was available. Michael Lowry from the National Air & Space Museum operated camera. [SVP photo, Brien Williams].

Betacam was the format of choice for all one-camera shoots as well as for two-camera shoots, when groups were interviewed onsite, rather than in a studio. The format was portable for fieldwork and provided excellent image and sound quality. Several two-camera shoots were edited live during taping, with the director selecting cameras images as they were recorded. At least one project would have worked better had the director been more attentive to environmental conditions and had prepared a more effective group arrangement, given the limitations of the site: continuous white noise, large center pillars (which limited camera placement and frustrated the crew), and a poorly designed table arrangement. Other two camera projects were not mixed (edited live) on site. This method gave the historians two distinct images of the session, usually a wide shot and a series of closeups, which allowed for more flexibility for later post-production editing. The historians who used this method preferred it because they wanted to make the final choice of images rather than letting the director select the shots at the time of recording. This technique requires post-production editing to combine all shots onto

One camera was used for discussion with three people at the Institute for Biomedical Problems in Moscow, Russia. The professional ability of the camera operator, Thom Wolf, allowed him to follow the flow of conversation while capturing a variety of effective shots. [SVP photo, Phillip R. Seitz].

one tape, and no historian has yet found the time, money, nor inclination to complete the process. Two cameras were usually the most flexible way to record a group interview, on site, without the extravagance of a full three- or four-camera shoot.

We used three cameras for group interviews with four or more participants to capture the widest range of facial expression, interaction, and reaction. These multiple camera sessions were always mixed on site; historians, therefore, trusted that the director selected the best camera shots to include on the final tape. Although we used three camera setups for a number of early projects, the more experience we gained regarding technical issues, and the videohistory process in general, the less we used this complex camera arrangement. Three cameras took more time for setting up and striking a set, required a good deal more equipment, involved upwards of seven or eight technical specialists, and usually encouraged excessive shots of the historian asking questions or nodding in agreement—reaction shots which were unnecessary in the final product. The extra camera was useful, however, for obtaining closeups of artifacts or photographs included in discussion. Our three-camera shoots were taped on open reel 1″ and on U-Matic formats.

DESIGNING A SET

As with selecting equipment and personnel, designing a set is important for capturing the goals of a videohistory session. Videohistory set design does not mean undertaking elaborate construction or seeking special decor to create a comfortable setting. Nor does it mean totally disregarding the environment in which the interview will take place.

For us, set design usually meant adapting to the given onsite location, a factory or office, for example. When faced with these environments, we worked the best we could with such ambient factors as lighting, sound, cluttered spaces, extra workers, squeaking copy machines, and overzealous museum curators, and sought to place the participant in the best place possible. In many cases, we wanted to record these unique sites, which meant that an experienced director and field crew were critical for such situations. They knew how to adapt to the environment to provide high level sound and lighting without distracting from the ambience.

When we were able to choose the location of the interview, such as a corporate board room or museum library for group discussions, we could usually monitor environmental conditions and place participants accordingly. If the location had little to do with the subject, we tried to keep the background very simple. If not, the result was often confusing. For example, we taped in a museum, with a variety of artifacts surrounding the interviewees and moderators. The artifacts, however, had little to do with the subject matter and distracted from both the conversation and from the objects which were relevant. An observant reviewer asked why the participants were located in such an odd place, with items that had so little to do with the conversation. For another project, participants were placed in front of a large nuclear reactor face, which was an appropriate backdrop for the topic of conversation. But the site dwarfed the people because the producer kept the cameras at a distance rather than shooting closer. Thus, the set design, and the inexperience of the producer and crew, negated what otherwise would have been a visually informative document.

Simple studio sets, with a plain background, were often more effective for group interviews when the objective was to document interaction and discussion, rather than objects and environments. Sites were used to enhance a topic rather than distract from it.

The placement of participants on the set was another crucial aspect of videohistory set design, particularly for group interviews. Effective arrangements encouraged discussion, captured interaction, and facilitated the presentation of artifacts. This was, however, not easy to achieve with groups over four. We tried several arrangements, but found that only one worked really well. A triangular table encouraged a more intimate setting by bringing participants face to face, and also provided a surface for placing objects. It allowed the historian to be present as an integral part of the interview without becoming a visual focal point. Although we had a triangle table built, one studio we worked with had a smaller one readily available. So, even though such tables may not be common, they can be found (or made).

Two video projects used the triangle table for group interviews with some success. Pam Henson, when conducting the paleontology group session, handed fossils to the scientists and operated a back-lit slide projector while the camera recorded the images. Stanley Goldberg, after several attempts at different arrangements, tried the triangle table for sessions taped in a studio. He was pleased with the way it brought the participants together, liked the simplicity of the set and its unlit background, and appreciated that the cameras were placed around the table to capture participants in enthusiastic discussion.[15]

We used a less formal arrangement for the California software designers. The six people sat on couches and chairs, crosslegged, with one person barefoot. Jon Eklund, the interviewer, sat between two cameras, off screen. This set worked well for this particular group because of their familiarity and youthful informality. They knew each other well enough to engage in immediate conversation, and were comfortable enough with the medium without being self-conscious or conspicuous.

We tried a similar arrangement with elderly technicians and scientists, who sat around a coffee table in a brightly lit makeshift studio while discussing their role in the space program. It didn't work. The moderator sat at one end of the coffee table, very predominant throughout the entire production, while the senior mentor sat at the other. The mentor's three subordinates sat alongside the table, obviously nervous as they fidgeted in their chairs. They deferred all questions to him. The design did little to facilitate discussion, and the dynamics were controlled by the presence of the mentor.

Less successful arrangements discouraged spontaneous conversation among participants, made handling artifacts awkward, and placed too much attention on the moderator. Some arrangements actually resembled sets for a political commentary television show or a government briefing room. Goldberg's early group interviews were designed as an open "V," where the participants were lined down

The triangle table used during Pamela Henson's Smithsonian Paleontology Videohistory Project gave her the flexibility to hand artifacts to participants and to show images from the backlit slide projector, to her right. These photos show initial setup, not the actual interview. [SVP photo, Terri A. Schorzman].

each side of a table while Goldberg sat at the open end. A camera was placed behind Goldberg, so all wide shots of the group included his back, which provided a strange image. Other producers chose variations of this arrangement, but placed the primary participant at the apex of the "V," so that he became the focal point, or opened the "V" so wide that participants on either end craned their necks to see each other. During a session where the open "V" was used, the moderators sat squarely in the middle, and conducted the sessions as though it was a weekly taping of a network political discussion program. For another project the producer moved people around after the first change of tape, which helped both the flow of conversation and the quality of the recorded images of the participants. Finally, another open "V" arrangement resulted in profile-only shots because of limitations of set design, camera placement, and technical difficulties. The producer, not wanting to disrupt the event, chose not to stop the taping to readjust for a more workable arrangement.

For a series of group discussions at a studio in Tucson, people were

The "V" arrangement of Stanley Goldberg's videohistory session at Oak Ridge, Tennessee for the Manhattan Project series. [SVP photo, James Hyder].

A "V" arrangement was used for the RAND Corporate History Videohistory Project at the RAND Corporation in Santa Monica, California. Historians were Martin Collins, far left, and Joseph Tatarewicz, far right. [SVP photo, Terri A. Schorzman].

placed in a large "V" (which had the appearance of a semicircle), without a table. David DeVorkin, the interviewer, introduced each session, then moved out of the scene. The scientists, although eager to talk about their role in developing an innovative telescope, would have been more relaxed had they been able to lean against a table. A table would have helped the use of documents during the interview as well. DeVorkin placed illustrations and articles on a small easel and passed them around; had there been a center table, it would have been much easier to handle and record closer shots of the items. A table might have made the set more personal by bringing the participants closer.

Other interview arrangements could have benefited from the use of a table. We found that people, particularly the elderly, preferred to have a table in front of them so they could set a glass a water and rest their arms. One of Goldberg's last group sessions was taped with a single camera. The placement of the four people, however, both was distracting to the viewer and caused the interviewees to lean forward to see each other while talking. Had they been placed in a small semi-circle around a table, they might have felt more comfortable in

The "V" arrangement was also used for the RAND Corporate History Videohistory Project that was taped at the National Air and Space Museum in
Washington, D.C. Collins, right, and Tatarewicz, left, were initially placed at
the apex of the "V" but were moved to this position. Artifacts were placed
on table in front. [SVP photo, Terri A. Schorzman].

front of the camera and the setting might have appeared more personal. Another similar arrangement also involved elderly scientists,
and took place in a planetarium's classroom where the producer had
little to work with in the way of set design. He had chairs, a brick
wall, long tables, chalk boards, doorways, windows, two cameras, and
an uncooperative crew. The result placed three men in a row, against
a plain brick background. The arrangement could have been more
appealing and comfortable by placing the men around the end of a
table. The director later noted that had he placed the men at the
end of a table they would have had something in front of them "to
make them feel more at ease and the angles among them might have
prompted more interaction."[16] He tried the arrangement on a later
session and was quite pleased. See Appendix 2 for a diagram on various group arrangements.

Arrangements for groups of two, or for one-on-one interviews were

Astronomers, opticians, and engineers gathered in Tucson to talk about their work on designing and building the Multiple Mirror Telescope. Had they been placed around a large table, documents and artifacts could have been incorporated more effectively. [SVP photo, Terri A. Schorzman].

less complex, since these interviews usually involved work with an artifact or touring a site. Site tours were usually conducted with a hand-held camera; occasionally the camera was placed on a dolly, or set on a stationary tripod for simple panning of a room. Nonetheless, we paid attention to lighting, sound, and image quality, and tried to avoid static talking head shots for any length of time.

THE COST OF VIDEOHISTORY

We tried neither to let video technology drive production requirements, nor allow ourselves to become intimidated by the cost of equipment needed to accomplish our goals. We had the flexibility and funding to use any combination we wished, and experimented with the medium by using a variety of formats, arrangements, and equipment options. This, however, is usually not the case for archival projects. The cost of video varies from high-end broadcast production standards to consumer grade camcorders. We rarely spent top

This group, recorded at the RAND Corporation, was placed in a single row. The table, however, made them more relaxed and provided a place for water, notepads, photographs, and artifacts. [SVP photo, Terri A. Schorzman].

dollar for elaborate systems and extremely high standards, but still found the average cost of professional systems to be prohibitive for most people or institutions interested in recording videotaped oral history, unless especially funded for the purpose.[17]

The following section is a guide to the budgeting process for both the cost of program administration and individual projects. Although we were generously funded, we set goals and determined how to accomplish each project. This meant spending time on the telephone with a variety of vendors, such as production houses and transcribers, to find the best prices.

We gathered approximate daily costs for hiring a crew; few will work a half day. A crew's fee includes equipment rental (or they will build in the estimated cost if they own their own) and professional time, and they will specify what, for them, constitutes an eight-hour day. For many, eight hours included travel to and from the site, plus setup and breakdown. Several agreed to work a ten hour day for eight before charging for overtime. Some crews included tape stock in their bid, but it was usually more expensive than if we bought tape directly from a vendor. Thus, we specified in our contract that we

Here, director Brien Williams placed participants in the Cell Sorter videohistory project at one corner of a table for better interaction. Historian Ramunas Kondratas is at left. [SVP photo, Joan M. Mathys].

provided tape. Finally, to our initial surprise, most crews expected snacks, a full lunch, and plenty of soft drinks. They rarely brought their own.

A one-camera professional betacam crew averaged $1,200 daily throughout the nation. We spent less than that, and we spent considerably more; we tried, however, to stay within the $1,200. A multiple camera studio shoot was upwards of $6,000 a day.[18]

We spent between $350 and $600 for the services of a director. Some worked for less money during the planning and post-production phases. Our "post-production" meant report writing, followup meetings, and reviewing finished products, not preparing a final, edited presentation. Since they usually hired the crew, we added the crew cost to the director's contract. Many preferred this arrangement because it placed the crew under their immediate direction, gave them more control over the product, and prevented the crew from answering to both the director and us. The arrangement was also effective when taping outside our locality. Many producers have a network that can recommend good crews, making selection and hiring much easier.

We included travel, duplication, transcription, and miscellaneous costs in each project budget. The SVP taped in a variety of cites and towns around the country, and in several locations overseas. As a result, the SVP spent almost $1,000 per traveler—historian, director, a staff member, often participants, and occasionally crew members— for each project. We included obvious modes of travel to the site, as well as other transportation requirements such as taxis, rental cars, subway fare, toll booths, and parking. Per diem costs were quite high, particularly in large cities, and weren't limited to production days only. We often arrived one or two days before taping—or took a completely separate trip—to survey the taping location, meet with site liaisons, local crew, and participants, and to conduct final research and gather objects as needed. We also needed to pay all expenses for time between sessions, which could amount to a day or two for projects taped over several days. Meals, hotels, laundry, and other necessities were included for daily expenses. We also included extra money to cover special meals and receptions for guests, crew lunch and snacks, and "presentation gifts" for when we traveled to Russia on two occasions (an apparent prerequisite).

We budgeted enough money to cover archival processing, such as tape duplication and transcription of the audio track. These procedures are addressed in the following section. We checked with local production houses regarding duplication fees (they vary), and sought bulk duplication rates when possible. We eventually used the services of an in-house duplication audio-visual facility, once it upgraded its equipment.

Transcription rates vary too, particularly when charged as an hourly rate rather than on the amount of tape recorded. For example, rates averaged $95 for each hour of recorded tape, not for time needed to transcribe one hour of tape. For example, a five hour video session at $95 per hour would cost $425. If it took thirty hours to transcribe the same session, at a rate of $25 (for example) per hour, the cost would be $875. A significant difference. We talked with a variety of transcribers about their fees and made sure we knew what we were paying for. Transcription procedures are discussed in the section on archival considerations.

Finally, we budgeted costs for preparing, or processing, the material for use. This represented staff time for final preparation of materials for deposit, which took approximately fifty hours for one person, per one hour of tape. Extensive follow-up research for technical terms, full names, or other citations resulted in significantly higher costs. We spent, on average, between $500 and $700 to "process" one hour of tape; this figure includes fees for outside duplication and transcription, and staff salaries (between $10 and $15 an hour) for final preparation of material and deposit in the SI Archives.

Miscellaneous budget costs include film and film processing for still photography, photocopying, long-distance and fax charges, audio cassettes and recorder, and social functions such as a reception or luncheon for the participants, historians, and crew. Appendix 2 contains a checklist of budgeting requirements.

ARCHIVAL CONSIDERATIONS

Our experience in producing, handling, processing, and preserving video collections was modeled on existing procedures in related fields, such as oral history programs and film archives. Our system was developed for access to original archival footage, since one of our major goals was to make videotapes and supporting materials available to researchers. To ensure both access and longevity, we followed the guidelines for processing and storing material as established by the Smithsonian Institution Archives and its Oral History Project (OHP),[19] but also incorporated elements that were unique to interpreting visual information in a narrative form.

Thorough archival processing, for both access and preservation, began during the shoot. If information was handled correctly at that time, our final steps were made easier. If not, archival post-production was both time-consuming and expensive. During SVP sessions, program staff always handled these tasks. The historian was concerned with content and participants and did not take on these important additional duties. In cases where a staff member was not present, due to limited travel funds or other circumstances, the historian and/or director attempted to complete the tasks. Inevitably, tape logs were incomplete, tapes were mislabeled or switched inadvertently in the field, and supporting field notes and observations were missing.

Before taping began we labeled master tapes with basic information such as date, name of participant, and session and tape number; we wrote on the non-stick labels located inside the sleeve of the cassette or used masking tape or some other removable means of identification on the tape itself. Temporary labels were removed for more permanent labeling at a later time.

While on site we also noted the interview location, subject information, and name of participant for each tape identified on a tape log (see Appendix 2). When possible, we also indicated the span of time code for each tape (provided by camera operator). This task enhanced the archival record and served several purposes. It provided general subject information about the shoot for reports, helped identify key words for the creation of a name and word list for transcription, and allowed access to specific segments of the interview before it was transcribed and cross-referenced.

The tape log was also an excellent way to verify correct spellings of names, places, artifacts, or technical terms with the participants immediately after the interview. Participants often spoke with a great deal of assumed knowledge when the cameras rolled and, as a result, referred to agencies or artifacts in slang or acronyms during the interview. We tried to define the terms before the interviewees left the site.

Another important task which eased archival processing was to collect resumes or short biographies from each participant. Resumes clarified dates of employment, listed publications, and provided other pertinent facts that were often overlooked during the interview. We initially asked the historian to request this information before a taping session, but we were rarely successful in obtaining it. Thus, we spent extra time tracking down the data long after the interview ended. In some instances a historian, or a site liaison, received biographical information ahead of time, which made archival processing more efficient.

At the completion of a shoot, we hand carried our master tapes from the taping site to the office. We never shipped them, even in overnight mail. The investment was too great to chance losing them in transit. To be extra safe, we handed them to security guards at the airport, rather than sending through X-ray machines.

Once in the office, we immediately attached yellow labels to both the tape and storage box to identify first generation master tapes. Our labels contained the following information:

[Sample]

SMITHSONIAN VIDEOHISTORY PROGRAM
Waltham Clock Company
Carlene Stephens
Session One: June 24, 1989
Tours and Interviews
Waltham, Mass.
20 minutes
Tape 2 of 3   1st Generation Master

Once the master tapes were labeled, we sent them for duplication. For the first three years of the program, we worked with a production house that provided us a low cost bulk rate, and had the capability to run all tape formats (1-inch, betacam, U-Matic, VHS). By the fourth year, the vendor no longer offered the bulk rate, so we turned to an in-house facility that had purchased more sophisticated equipment.

We made second generation tapes, which were dubbed from the master tapes to U-Matic 3/4 inch; these became the dubbing masters, from which any third generation VHS copies were made. We used blue labels for U-Matic, and white for VHS, thereby making it easier for the SI Archives reference staff to identify tape generation and not accidently play a master tape or loan a dubbing master.

Time code was transferred at the time of duplication as an electronic signal from the master to the U-Matic, so that copies could be made with or without the viewable time code "windows." We specified, however, that all research copies contain the window. This was both a way to facilitate research and to prevent mass duplication without written permission, our assumption being that not many people would copy tape with burned-in numbers unless it was needed for a specific reason.

While tapes were being duplicated and labeled, we asked historians to prepare a name and word list to assist transcription. Since much of our work was very technical and scientific, we required a detailed list that included proper spellings of personal names, institutions, and titles, and a full description of acronyms, technical jargon, locations of companies, and other references. This information was critical for the preparation of an accurate oral transcription; most transcribers preferred that the list follow the order in which the terms were used, rather than be arranged alphabetically. If a good subject tape log was created on the day of the shoot, it assisted historians with preparation of a full list. Some historians had taken notes during the interview and could add to our efforts quite completely. Others asked that they receive a copy of the tapes to review the session while they prepared a list. Most, however, were fully aware of this aspect of the project before undertaking the work, yet did not comply with this "post-production" request. They were either too busy, or were simply not interested in the archival process. As a result, we prepared name lists and asked that historians review them before transcription. In any case, a good name and word list not only helped the transcriber, but made editing much easier.

Although expensive, we transcribed the audio track of all videohistory projects since historians are accustomed to working with the written word; thus, audio narration and visual evidence were available to researchers in written form. Transcripts, however, were not intended to replace viewing the tapes. The SIA's Oral History Project (OHP) manual cites excellent reasons for transcription, including 1) a researcher can read three times faster than listen or watch, and is thus more likely to use the tape if it is transcribed; 2) the transcript spells and verifies names, dates, and details; 3) the transcript reduces the threat of misquote and thereby reduces the debate over what was actually stated; and 4) the transcript can contain photographs, footnotes, reproduced maps and correspondence, and any other documentation that could add to the session. Thus, a transcript contains additional data for a researcher.

The OHP manual also notes that an interviewee can review a transcript and can finish lines of thought, add details, and clarify topics. We, however, did not ask participants to review the transcript in draft form, due to the considerable amount of time needed to do so. We interviewed far too many participants (over three hundred) in a relatively short amount of time to ensure that all transcripts would be returned promptly—if at all. Instead, we lightly edited transcripts and provided correct spellings and full citations for people, institutions, and technology mentioned in the interview. We then asked historians to ensure the document's accuracy by performing additional research as needed or by talking with participants. They usually reviewed our comments and answered immediate questions; they spent less time on extra research or discussion with participants.

Videohistory transcripts presented us with the dilemma of how to gain access to visual evidence in the narrative form. Our solution was to include time codes and abstracts of visual information (cues) on the transcript when editing as a way to verbally describe visual evidence. Cross-referencing the time code from the tape to the transcript brought together the two forms of historical documentation. Time code notations allowed a researcher to scan a transcript and then go directly to places of interest on the tape. Most SVP projects consisted of multiple sessions and averaged six hours in length, so the time code gave researchers immediate access. It also allowed scholars to cite the specific visual image as a footnote in a book or article. Time codes generally accompanied a visual cue on a transcript, or, in the absence of significant action occurring in the interview, were used alone to document subject changes or simple evolution of time. If a subject did not change, nor was an object introduced, we placed codes every couple of minutes. An example of a time coded document would look like:

02:35:44:00
SMITH: Please show me how the vacuum testing chamber
                works. It seems to be a complicated piece of ma-
                chinery.

JONES: O.K. First you lift the lever, then you pull. . . .

Visual cues describe action that occurs on tape which cannot be
understood by reading the oral transcription, such as when an interviewee's transcribed dialogue does not make sense without the ac-
companying description of an action. In order to keep the visual cues
as brief as possible (usually no more than two lines) we used short,
direct statements and active verbs, such as:

00:32:44:00   [Cooper adjusts camera height]
or
03:44:01:00   [Goldberg hands wrench to Bainbridge]

These cues provided important information when reviewing the transcripts, but are not substituted for viewing the tape.
    Time code and cues also work for videohistory tapes that are void
of oral narrative, but are rich with visual images. To document this
aspect of a tape, we wrote a paragraph that described the action, or
listed the artifacts, within a span of time codes, such as:

00:01:05:00   [Conveyor carries large pieces of slate from quarry
                to cutting room].

00:05:02:00   [Beayon slices one-inch thick pieces of slate from
                large block, tosses inferior pieces into scrap pile, and
                sends quality slate down conveyor for precision cut-
                ting].

This method enabled us to create a textual document that records
purely visual information. When we shot additional closeups to record artifacts, we added page numbers to reference where the item
appeared in the interview, so a researcher can view the artifact within
context, such as:

[B-Roll: Photographs]

01:33:45:00   Janet Bragg in aviator's uniform, c. 1932. [p. 20]

01:33:46:00   Alfred Anderson in cockpit of plane, c. 1940. [p. 32]

We also indicated the location of an interview in a transcript, particularly when the site changed several times *within a session*. For example, a session at an observatory included interviews at a number of different locations, each selected to convey or discuss structural and design features of the telescope. Since the place of the interview was central to discussion, we included it, along with the name of the participant(s) each time it changed.

02:22:10:00

LOCATION TWO:

MMT: Telescope Observing Chamber
Fred Chaffee, Craig Foltz, Carol Heller
May 9, 1989

If a session included a lengthy tour of a site's facilities within a general area, the number of location changes would be impossible to note in separate headings and were included within a visual cue, such as:

00:25:50:00   [Camera follows Genin through laboratory and specimen collections at the Biomedical Institute]

Finally, some sessions, as mentioned in Chapter Three, were shot unmixed with two cameras in order to record separate images of the interview. This technique, while providing a good deal of visual information, presented yet another special problem for transcribing: conveying two distinct actions in a video interview, even though oral narration is the same. To handle this situation, we discussed the special feature on the first page of the transcript, and then noted images from both cameras in the cue, as:

00:50:42:00   [Camera A: Close-up of Comstock shaking her head no in response to a question]
[Camera B: Two shot of R. and K. Williams nodding in agreement]

Transcription for videohistory, then, involved a good deal more than a verbatim narrative of oral discourse. Although time-consuming, a well-prepared transcript, when combined with a tape, is an extremely useful resource. That resource is even more effective when used with a finding aid. A finding aid describes the content and context of a video session, such as historical background of the topic, reasons for undertaking the project, biographical data about the participants, and a summary of both narrative and visual information. We wrote one finding aid per project, whether it consisted of one session or twenty. The finding aid was bound with the transcripts, to provide better access to material for researchers. A fully bound transcript includes: title page, copyright page, program description, release forms, finding aid, name list, text, general index, and processing log (provenance). See Appendix 2 for an example of a finding aid.

Finally, archival processing of videotape includes the important need for long-term preservation. We considered it at the outset of the SVP, not at its completion. Video is not an archival medium; the information can be altered or removed from the original, and the shelf lie is not guaranteed. Our strategy was to implement steps to safeguard the product.

All presently known electronic media are inherently unstable. Even the best quality master tapes kept under ideal environmental conditions and never played have an expected shelf life of no more than twenty years. Those that last longer will contain noticeably deteriorated information. This is a real concern for archival projects which record, collect, or store videotape.

Although they are quite useful and stable for time periods of ten to twenty years, current flexible magnetic recording media suffer from recognized material degradation processes that make them vulnerable to large-scale information losses over long periods of time. Their advantages of rapid access times and adaptability to large-scale data manipulation are not particularly relevant to archival needs. Their ability to be erased or updated is, in fact, a liability in archives. Material degradation processes in current optical disks are not well understood, and the technology is not sufficiently developed to recommend them for archival use.[20]

To hedge against loss, store master tapes in an environmentally controlled room at 65 to 70 degrees and 35 to 45 percent relative humidity (avoid fluctuation in temperature), keep them on dust-free steel shelving with a baked enamel finish, place them vertically, like books, to avoid putting stress on the tape, and keep them in fire retardant containers, not cardboard boxes often used to store tape.[21] Also, since tapes degrade through constant playback (playing seventy-five to one-hundred times will result in significant image loss due to flaking of magnetic impression), the SVP plays master tapes only once, for the purpose of creating dubbing masters. Master tapes will eventually be housed at an off-site facility, in cold storage. In addition, we recorded on the best tape with the best equipment possible (higher quality tape, such as 1″ or high-speed 1/2″ Betacam is much better than consumer grade low-speed 1/2″ VHS), duplicated the master, and kept masters and subsequent copies separate. We plan to remaster to new formats when feasible (financially and technically) so images can be retrieved as equipment gets better.

Magnetic tape and optical digital disk playback and recording equipment change so frequently that they become vulnerable to obsolescence, which "could render data unreadable even if preserved on the primary medium."[22] Because of this, the Smithsonian Archives will maintain equipment to ensure access to recorded contents and eventually remaster, will keep currently operating equipment clean and demagnetized since magnetized heads may alter the recorded signal, and will retain it when new equipment is purchased, for the purpose of playing older formats.[23]

SUMMARY

Recording videohistory brought together a team of professionals that resulted in working relationships different from those established in traditional historical research projects. Historians provided the content and worked closely with technical experts who provided expertise in video production. The SVP staff gave administrative and project support.

The production team selected recording technology according to individual project requirements, balanced the cost of equipment against expected quality, and tried not to let technology drive the project. Recording videohistory also required a long-term commitment to archival processing, such as duplication and transcription, and to maintenance of the collection, such as provision of proper storage facilities.

Evaluation of the product by those who worked closely with it follows.

NOTES

1. SVP, "Annual Report: Summary of First Year Activities and Request for Second Year Funding (FY88)," September 1987, p. 12.
2. The director handled technical arrangements and supervised the crew;

the crew included camera and sound operators, and occasionally assistants such as gaffers; the producer was responsible for intellectual content and project overview, and, in our case, was the historian (although many directors were also called producer).

3. Brien Williams, Report on Paleontology Videohistory Project, June 1987. Also, see Williams' article in Chapter Four.

4. Brien Williams, Report on the Multiple Mirror Telescope Videohistory Project, July 3, 1989.

5. SVP, "Annual Report for Second Year Activities and Request for Third Year Funding," September 1988, p. 21.

6. Selma Thomas, Report on the Waltham Clock Company Videohistory Project, July 27, 1989.

7. Michael Williams, "Review of the J. Presper Eckert Interview Video Tapes," p. 10.

8. Ibid.

9. Terri Schorzman, "Status Report to Robert Hoffmann, SI Assistant Secretary for Research" July 12, 1990; Schorzman, *Technology and Culture*, January 1989.

10. For good introductory references on video technology, see David Mould, "Composing Visual Images for the Oral History Interview," *International Journal of Oral History*, 7 (Nov. 1986): 198–205; Brad Jolly, *Videotaping Local History* (Nashville, TN) 1983; Bruce Jackson, *Fieldwork* (Urbana, 1987), Brian Winston and Julia Keydel, *Working with Video: A Comprehensive Guide to the World of Video Production*, (London, 1986). For technical terms, see Glossary, Rick Prelinger, *Footage 89: North American Film and Video Sources* (New York, 1989).

11. Comment from John Fleckner on his review of draft of this book, October 31, 1991.

12. For further information on time code, see "Introduction to Time Code" in the appendix. Also, EECO, Inc., *SMPTE/EBU Longitudinal and Vertical Interval Time Code, 1982*.

13. See sample contract in appendix 2 for example of what to specify for a crew.

14. David DeVorkin, Report on Classical Observation Videohistory Project, March, 1988.

15. Goldberg addresses these issues in Chapter Four.

16. Williams, Report on the Multiple Mirror Telescope Videohistory Project, July 3, 1989.

17. Alternatives to renting high-end systems include hiring students from a university or community college media/communications department, or becoming involved with the public access cable channels, which often offer equipment free of charge. See "Public Access Television Opens New Era in Oral History," *Michigan Oral History Council* (Winter 1989), 2, 7 for an analysis of using public access to record videotaped oral history interviews.

18. To purchase systems: A consumer grade VHS or VHS-C camcorder will cost at least $700; a higher grade S-VHS or Hi-8 system will vary between $1,500 and $2,000; a low grade industrial averages $6,000; and a more professional system will cost at least $15,000. Local sellers will determine the current rates. See the "Guide to Gear" issue of *Consumer Reports* (March 1991). For information on general production techniques, see Jackson, *Fieldwork*; Mould, "Composing Visual Images; Jolly, *Videotaping Local History*; Winston and Keydel, *Working with Video: A Comprehensive Guide to the World of Video Production*; Ronald Compesi and Ronald E. Sherriffs, *Small Format Television Production* (Boston, 1985); Herbert Zettl, *Television Production Handbook* (Belmont, Calif., 1984).

19. Pamela Henson, "Oral History Project Manual," SIA, for full description of SIA processing. Also, Joan M. Mathys discussed SVP archival procedures during her presentation at "Videotaping Oral History," a four-day workshop in New Zealand. Much of this section is based on her formulation of videohistory processing techniques.

20. *Preservation of Historical Records*, Committee on Preservation of Historical Materials of the National Materials Advisory Board, Commission on Engineering and Technical systems of the National Research Council, National Academy Press, Washington, D.C., 1986, pp. 85–86.

21. Alan Lewis, "Fact Sheet on Videotape Preservation," January 1991.

22. Preservation of Historical Records, pp. 85–86.

23. For other sources on archival preservation see: Dierdre Boyle, "Video Preservation: Insuring the Future of the Past," *The Independent* (October 1991), p. 25–31; "Film/Videotape Factsheet," *Conservation Administration News* 22 (no date); "Film and Videotape Preservation Factsheet," *Footage 89: North American Film and Video Sources* (New York, 1989), p. A-28, A-29; Randy Hayman, "Archiving Videotape," *AudioVisual Communications* (March 1991), p. 21–25; Roma Kemp, "How Long is Forever?" and New Methods of Data Storage," *Northwest Oral History Association* (Fall/Winter 1990), pp. 2, 4–7; Rick Prelinger, "Archival Survival: The Fundamentals of Using Film Archives and Stock Footage Libraries," *The Independent* (October 1991), pp. 20–24; "A Roundtable on Television Preservation," Rapporteur Summary of a Colloquium Convened by The Annenberg Washington Program, (May 19, 1989); David Shulman, "Deja Vu—Resorting and Remastering Open-Reel Videotapes," *The Independent* (October 1991), pp. 32–35; 3-M, "The Handling and Storage of Magnetic Recording Tape," *Retentivity*, no date; 3-M, "Videocassette Tape Physical Damage," *Retentivity*, no date.

4

THE EVALUATION
OF VIDEOHISTORY
IN HISTORICAL RESEARCH:
PERSPECTIVES OF PRACTITIONERS

INTRODUCTION
David DeVorkin And William Moss

The adage "a picture is worth a thousand words" is just not true. A picture, especially a moving picture, is different from a literal narrative. The two cannot be equated. Visual communication and comprehension is radically different from the dominant mode of academic discourse, i.e., words and linear logic. To evaluate the distinctive contribution of videohistory as a research tool, historians often need to translate moving pictures into analytic text, since text is the primary form of peer review.

A similar issue arose in the early days of oral history, when oral historians talked of a hypothetical "oral book" that would somehow be quite different from the linear tradition of reading information. The "oral book" never became a reality, although there have been dramatic readings, voice-overs from interviews that accompany slides, and enacted skits based on oral recordings that attempted to convey the original spirit. The dominant characteristic of oral history, however, has been words. Audio recordings have been transcribed into written transcripts, despite attempts to use original sound recordings as the primary source of evidence and evaluation. Nonetheless, transcribed oral history has been applied successfully in historical research and writing, despite early skepticism. Oral history has been accepted on terms of traditional documentation, but historians have still not come to grips with the unique contributions of audio evidence.

Some oral historians have embraced video recording as an available technology to extend their craft. They have noted that video is useful to capture environment, process, and interaction among groups of workers. The problem remained, however, of how to make clear in written language the visual contribution of video, and also how to rate it in comparison with other forms of evidence that are more easily measured or accepted. We can approximate. This means that the controlled use of video as a contribution to history is likely to remain experimental for some time to come. How video is used in historical writing, or even in the production of video documentaries, is likely to remain an open question that will be resolved by cumulative experience and evolving practice, not prescribed methods.

We hope that the Smithsonian experiment moved this process further along. Since the Smithsonian's goal was to generate video materials driven by the research requirements of SI historians, we intended that the historians full immersion in the experience would provide a reflective assessment of the contributions of the recording medium. The historians contributing to this chapter were asked to provide their impressions of how well the Smithsonian Videohistory Program served their individual research goals. They were selected deliberately because of different approaches and perspectives. Their essays review the set of goals that most of the participants shared with the Smithsonian administration and the Sloan Foundation at the outset of the program. One goal was to test all accessible techniques of video interviewing, as Stanley Goldberg's essay makes clear, based upon his extensive series of taping sessions with members of the Manhattan Project.

Another goal was to focus on how video might augment appreciation of tools and devices in the national collection by encountering and preserving what remains of the shop practices that either created these objects, or used these objects as tools to create other objects. This was the intent of Carlene Stephens in her study of the Waltham Clock Company. A third avenue, pursued by Steven Lubar, was to document contemporary events and modern laboratory processes. He indicates in his evaluation of contemporary robotics efforts how and why he feels this is a preferred use for video documentation, in contrast to using the medium to capture old processes and events in the historical memory of participants and observers. And fourth, David DeVorkin reflects the SVP goal of determining how videodocumentation might be used to augment the on-going historical research of Smithsonian staff, to the point of its incorporation into a research paper. He examines a group active in early space science research.

The essays offer impressions of experiences with video, and, to varying extent, reflections on the process of preparing for and conducting the interviews. The sum of these statements provides a balanced overview of the typical interests and priorities all of the participating historians brought to the enterprise.

A fifth essay, by Brien Williams, demonstrates the degree to which video production remains highly technical territory, within which craft, forethought, and handwork are as evidently important as they are in the Waltham watchmaker's world, so nicely encapsulated in Stephens's essay. The sixth paper, by Kerric Harvey, who is from the world of communications research and applications, is useful in that it reveals the potential range of interpretation that can be applied to the visual material gathered during the program.

The videohistorian then, if such a person is to ever exist, must embody not only Williams's or Goldberg's sense of craft, but Stephens's and Lubar's sense of historical mission. This became clear to us after the second year of the program, as we collected external evaluations from those in fields knowledgeable both in craft and content in the areas we addressed.[1]

When the combination worked, it created "a texture and flavor" which, as one reviewer noted, "other sources simply cannot match."[2] Historical figures who had existed to the historian only as writers of journal articles or creators of institutions and techniques now could be seen, if not in action, at least in reaction to the interviewer's queries or requests for demonstrations, or to a producer's request to perform an "enactment."

As most reviewers pointed out, videohistory documentation must adhere to high technical and intellectual standards if it is to be useful. The product has to be as focused as possible to preserve useful visual information. This goes far beyond image focus or camera stability; it requires choices of framing and composition, in effect, a connoisseur's selectivity. If the lens fails to capture the significant "action," future historians who try to use this material will be chasing shadows.

While each reviewer saw something different in the materials being reviewed, most were sensitive to both content and craft. We asked non-affiliated SI reviewers to comment on the "contribution these video interviews make to the history of [subject matter]; the contributions that the visual dimension makes to enhancing understanding or add to new knowledge; the methodological proficiency of the interviewer; and the technical quality of the video recording in capturing what was placed before it."[3] Although these questions seemed to "prime" the responses, the degree to which content reviewers were concerned with technical quality is significant. No reviewer argued that technical quality was not important. This indicates that, as yet, the medium of video is far from transparent; it very much influences the message it contains.

NOTES

1. Nine reviewers reviewed nine different projects in the autumn of 1988; project historians had recommended scholars from their research areas. Reviewers were paid an honorarium to do so. In January 1989, several of these reviewers, along with technical experts and SI administration, convened to review and assess video footage, comment and evaluate on videohistory methodology, and make recommendations for a permanent videohistory presence at the Institution.
2. Memo from Stewart Leslie to Terri Schorzman, October 26, 1988, regarding evaluation of RAND videohistory shoot.
3. Letter sent to all reviewers, 1989.

THE MANHATTAN PROJECT SERIES
Stanley Goldberg

It was in the summer of 1986 that I first heard that the Sloan Foundation was considering funding the Smithsonian in an experimental videohistory program. I was then a consultant to the Smithsonian's National Museum of American History and was in the midst of preparing a planning document for a large permanent exhibit on the history of the Manhattan Project. In that context I had been examining various archival holdings with regard to the Manhattan Project. My curiosity was piqued by the realization that the documentary record I was looking at suggested a perspective on aspects of the Project which was, as far as I could tell, nowhere reflected in the secondary literature with which I was familiar.[1] And so when I finished my work on the proposal, I naturally turned to a host of questions which my initial research had spawned. I was particularly intrigued by four questions: The conditions under which the Project was started, the role played by General Leslie R. Groves in the decision making process, the dynamics behind the decision to use the bomb, and the initial postwar posture of the United States with regard to international control of nuclear technologies. These interests all coalesced into a decision to write a full scale biography of General Groves. Simultaneously the Smithsonian Videohistory Program committee asked me to submit a project proposal as part of the package to be considered for Sloan Foundation funding.

My initial work on the history of the Manhattan Project carried me to many of the original sites. I was impressed by the number of such installations which still existed and by the number of Manhattan Project veterans still living in those locations and working at the sites. Well-known Project scientists and engineers had been interviewed many, many times, usually for the production of one or more of the countless film and television documentaries on various aspects of the Manhattan Project. The immediate issue was the justification for doing yet another set of interviews on the Manhattan Project.

In a one-on-one interview, there is no question that being able to *see* the response of the individual can add information to the user; however, if one insists, as the Sloan and Smithsonian officials did in the name of archival preservation, that the recording be of the highest broadcast standards, then the cost differential between an audio interview and a video interview did not, in my opinion, justify getting the added visual information. I concluded that there were two types of interviews which might justify the added cost of video since it would make possible things that could not be done with audio alone: the first was group interviews and the second was the documentation of sites and artifacts.

The advantages of video for a group interview are fairly obvious. First, the medium itself provides the viewer with the identity of the speaker. But more than that, it is possible to document subtle visual reactions of participants. There had been very few, if any, group interviews with Manhattan Project veterans.

While there have been many photographs published of Manhattan Project sites, for the most part the details of what was done at these sites remained a public mystery. Initially, this was due in part to the demands of security classification. Later, after most such restrictions had been removed, there was no demand on the part of broadcast television to enter the previously restricted sites. While of historical interest, they were no longer news. Thus, detailed documentation of the sites on film or video was largely absent. The proposed Smithsonian Videohistory Program seemed to me a wonderful opportunity to fill that void.

And so it was on the basis of those two pillars—group interviews and the documentation of sites and artifacts—that the proposal was accepted. I intended to interview many of the leaders of the project, as well as groups of people from whom we had heard little, those who had worked in the factories without any knowledge of the purpose of their efforts. I saw this as a route to providing some insight into the social history of the Project and as a way of getting behind the standard stories from those who had been interviewed over and over.

Early conferences of the Smithsonian Videohistory Program advisory committee with principal investigators emphasized the experimental nature of the program. While planning for the Manhattan Project sessions, I made the experimental nature of our work an explicit criterion. When working with a video producer, we began with many questions about the procedures, such as the best approach to seating arrangements, the number of cameras to be used, or whether to mix live or to let each camera record the entire session. There were questions about interviewing technique and preparation: To what degree should participants be provided with an outline of topics? Should they be provided with copies of archival documents bearing on their experience and contributions? How should the session be introduced? How tightly should the moderator control the direction of the interview? There are no right and wrong answers to most of these questions. There is no one best way to conduct an interview. There is no best seating arrangement, no ultimate answer to the question of the relationship between cameras, no best way to conduct the interview. Everything depends on circumstance—personnel, environment, the rapport between participants and with the moderator.

The bedrock criterion which guided me in making decisions was the archival nature of the project. We were not engaged in a made-for-TV "production." Footage was not being prepared with a particular program in mind nor to make a polemical point. My purpose was to provide information on artifacts and sites, and on techniques of research and of manufacture and to make it available for anyone interested in using the material. I also wanted to allow those who actually participated in a historic event to recall, as best they could, what they thought they were doing at the time. It was also an opportunity for the participants to bear witness and, given the group nature of the interviews, to play off each others' memories. I hoped that these interviews would be somewhat self-corrective; that participants, reacting to each other's memories of events long past, would recall things not otherwise remembered. They might realize that certain sequences could not have been at the time nor in the order in which they remembered. Of course, it might also turn out that the stronger personality of a group might convince others and that his or her narration, even if improbable or downright incorrect, would prevail. As it turned out, both of these scenarios actually happened in several of the interviews.

GROUP INTERVIEWS

Site Selection

When planning for the Manhattan Project videohistory series, I was not sure just how many sessions would be possible. The Project itself formally went out of existence at midnight, December 31, 1946, but for all practical purposes the bombing of Hiroshima on August 6, 1945, of Nagasaki on August 9, 1945, and the surrender of Japan on August 14, 1945 marked the effective end of the Project. In planning the series, we were handicapped by the fact that we were dealing with events that had taken place more than forty years in the past. Many of the participants were no longer alive. Some of the sites were either gone or, like Los Alamos, had been so transformed as to have little or nothing to do with their wartime history. Though these fac-

Kenneth Bainbridge, left, and Robert Wilson, top of hill, were interviewed at the Trinity site, Alamogordo, New Mexico, during the Manhattan Project videohistory series. [SVP photo, Phillip R. Seitz].

tors defined the boundary conditions, in planning the sessions such limitations did not play any major role.

I wanted to visit certain sites for the purpose of documenting the installation and the artifacts located there. I realized, though, that regardless of historic sites, individual Manhattan Project veterans were concentrated in particular regions of the country. The following site locations quickly became a priority: Richland, Washington, site of the Hanford reactors; Oak Ridge, Tennessee, home of the major efforts at Uranium isotope separation; Cambridge, Massachusetts, a locale where many of the physicists and chemists who had been in leadership roles in the Project now reside; and Los Alamos, New Mexico, not only because it was home to Manhattan Project scientists and engineers but also because it was near Alamogordo, site of the test of the first atomic bomb in July, 1945. Also, since the Smithsonian possesses the *Enola Gay*, the airplane that dropped the atomic bomb on Hiroshima, as well as exemplars of the bomb casings for Hiroshima and Nagasaki, we planned sessions in Washington, D.C. that were organized around these artifacts.

Stanley Golderg, right, interviewed Norman Ramsey and Harold Agnew at the Smithsonian's Garber facility, where the *Enola Gay* is being restored. [SVP photo, Alexander B. Magoun].

Session Topics and Participant Selection

Having tentatively identified possible sites, I next defined topics that would explore the administrative and operational hierarchy of the Project that would not repeat that which had already been done. From the beginning I organized particular sessions with definite topics in mind. In some cases the topic was nothing more than exploration of artifacts or site installations. In other cases I had in mind a definite orientation, for example, "Women at Los Alamos," "Alamogordo and Tinian," "Electromagnetic Separation," "Oak Ridge Culture."

The choosing of particular participants represented a compromise between the ideal and the possible which sometimes redefined the actual topics discussed. I expended an enormous amount of effort identifying the availability of potential participants. I had two lists: the first contained the names of specific individuals, the second merely the type of person I wanted, such as Hanford reactor workmen, or chemists who worked in the Hanford separation plants, or Oak Ridge calutron operators. In order to transform the second list into a set of specific individuals, I relied on Manhattan Project veterans who had leadership roles in the Project. They were helpful in identifying and locating peers who could participate with them in group discussions. I also relied on public information officers at Manhattan Project sites still in operation for identifying potential interviewees who played supporting roles in the Manhattan Project. They led me to people who could talk about the daily social and cultural ambiance. This worked fairly well as long as the participants were identified in response to criteria which I had arrived at based on the topics I wanted to explore. When, on the other hand, the session was defined by the expediency of who was available, things did not work out as well.

A perfect example was a session at Hanford which featured several of the individuals responsible for the administration of the site during and immediately after the war. The Hanford interviews, which were done during the second week in January, 1987, were the first sessions in the series. They were scheduled to take advantage of a program of public lectures featuring Colonel Franklin T. Matthias, administrative head of the site during the war, along with others high in the Hanford hierarchy. These individuals also participated in a group interview centered on the administration of Hanford. It was my first group interview (as opposed to site documentation) and my sense of the session was that, except for Colonel Matthias, the participants were bewildered as to why they were being interviewed. As a result of that experience, I steadfastly refused sessions organized solely because the participants happened to be there.

On the other hand, I had to be prepared to take advantage of the serendipitous moment. Prior to going to Hanford I made careful arrangements to interview specific individuals who had been employed to maintain the first production reactor. When my producer and I arrived, we discovered a retired reactor operating engineer at the site who came to observe the session. After talking to him for a moment or two I realized that he was an ideal person to explain the nature of the various controls in the reactor control room. He was superb.

Toward the end of the series, I realized, in hindsight, that the success of any session depended most significantly on a variable beyond my control, namely, the chemistry that developed among the participants and between the participants and myself. This rapport did not always depend on how well the participants knew each other. For example, several sessions in Cambridge featured people who had not only worked closely with each other during the war, but who had kept in close contact since then. Some sessions went extremely well and others just flagged. On the other hand, sessions in which people had not seen each other for years or had never known each other moved along with a surprising and delightful pace and depth. My own hypothesis for this lies in the degree to which the participants identified themselves with the outcome of the Manhattan Project. Most of the scientific leaders of the Project, along with those who attained prominence in the fields of academic physics and chemistry, were cautious and tentative. They told a standard story concerning their participation in the Project and their later attitudes toward that participation. They maintained their story regardless of new information emerging from the video session. Some divorced their participation in the building of the bomb from their later career, as well as their later attitudes toward cooperation with military projects. Others aggressively defended their actions during the war without apology. In the same session, these two extremes often talked past each other and the result was less than satisfactory. As will be discussed later, my efforts to cut through these stances did not meet with great success.

This kind of session was in sharp contrast to sessions composed of people whose participation in the Project was only contingent to the rest of their lives. One such session was composed of four women who went to Los Alamos because of their husbands' work. Two had found scientific work within the Project and the other two had not. The discussion covered many topics, and issues of ethics and morality took on an immediacy which was largely absent from sessions in which male scientists discussed life and work at Los Alamos.

A second such session was composed of men and women who had been recruited during the war to operate the isotope separation machinery at Oak Ridge, or to train those who were to operate that machinery. They had no idea of what they were working on, and the announcement of the atomic bomb at the end of the war came as a complete surprise. There was a freshness to the discussion which is palpable on the tape and which is not present in the sessions recorded with the scientists who designed, installed, and maintained the machinery which these individuals operated.

This points up the fact that the ambiance of any one of the sessions had less to do with my approach than it had to do with the participants' own stake in the Project itself and with how openly they confronted and interacted with others on the panel.

The relative status of the participants was another dimension to be considered when putting together a group session. When the participants perceived themselves to be more or less of equal professional status, sessions were more lively. But, as might be expected, when the group was composed of a strong leader and those who had worked under him, deference to the leader of the group dampened the discussion. I am not sure that there is much that can be done about that situation. One strategy that I did not try in organizing such groups was to omit the person who had served as the team leader. It's an idea worth trying.

Organizing the sessions was an extraordinary experience. As I mentioned earlier, in the interests of economy, we selected sites so that we could conduct a series of interviews in a concentrated manner over a few days. Inevitably, this meant transporting several people great distances, and making arrangements for lodging and meals. It also meant finding local, suitable technical production people, arranging schedules, and locating an appropriate setting for the interviews. Making this happen was not a trivial undertaking. Our first attempt for the Hanford sessions presented seemingly unsolvable conflicts and unmanageable barriers. We somehow stumbled and staggered past the obstacles and in the process learned a great deal about how *not* to do things. We were blessed with superb support from the Smithsonian Videohistory Program office. All of this was as new to them as it was to me and the measure of how quickly they caught on was demonstrated by the relative ease of our second set of sessions (at Oak Ridge, Tennessee) even though the logistics were much more complicated.

Physical Arrangements and Production Values

Sessions that were not intended to document sites and artifacts, but to conduct group interviews, were done in professional production studios. This was the case at Hanford, Oak Ridge, and Cambridge. We used three cameras located roughly around the sides of an equilateral triangle. In earlier sessions two interview tables were arranged in a V-shape and I sat at the open end of the V with an easel within easy reach for the purpose of displaying still photographs which could be simultaneously recorded by one of the cameras and seen by the participants. In Cambridge, the two tables were replaced with a single triangular table at which we all sat. The general arrangement is depicted in the diagram.

Sessions were mixed live at the discretion of the producer, James

At the video shoot at Oak Ridge, Tennessee, the "V"-shaped arrangement included the use of an easel. [SVP photo, Phillip Seitz].

Hyder. The final product, then, was a single set of tapes containing the mixed images and the audio track. In these tapes the V-shaped arrangement showed long shots of all participants facing the camera and me with my back to the audience. James Hyder would often select shots showing individual reactions to particular points, including, from time to time, my reaction to what was being said.

Selma Thomas was the producer for the sessions at Los Alamos and Washington. For these sessions, we abandoned the V-shaped arrangement and the use of a production studio. Only two cameras were used and each of the cameras recorded different perspectives of the entire interview, the audio being shared between them. In this arrangement, though I can be heard asking questions and comment-

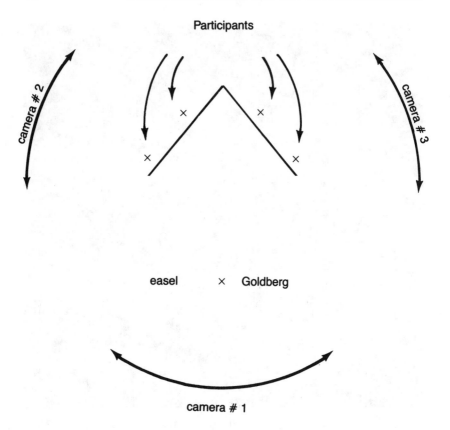

ing on the discussion, I never appear. On balance, while the earlier format allowed for a freer manipulation of images, strictly speaking it is at odds with the spirit of making an archival record and for that reason I recommend the format which avoids live mixing. On the other hand the use of a table for the participants is strongly recommended. In the Washington session, group interviews were done without a table, all the participants seated in chairs in front of the camera. Some of the interviewees told me, privately afterwards, that they felt uncomfortable with that arrangement. I worried that the fact that they did not face each other would tend to suppress interactions between participants. As it turned out, the four participants were anything but reticent. Nonetheless, it should be a consideration in planning such sessions. The arrangement also made for somewhat

A table, and a different arrangement, might have helped these participants seem more comfortable during a session recorded at the National Museum of American History. [SVP photo, Joan M. Mathys].

busier camera movement than had been the case in other sessions, which I personally found distracting.

Interview Style

As stated earlier, at the initial planning meetings of principal investigators and the steering committee of the Videohistory Program, we emphasized the experimental nature of the Smithsonian's videohistory projects. The techniques that one might use in a one-to-one audio interview might not be applicable to a group interview. We were not clear what would work, but, as I commented earlier, one of the most important factors was the rapport among the participants and between the participants and the interviewer. Still, in the spirit of exploring what was possible, I was determined to try a number of different techniques. Two other factors played an important role in shaping later interviews. The first was my experience from previous sessions, and the second was the change from live mixing and the concomitant change in the physical arrangement of the setting (See "Physical Arrangements and Production Values," above).

The experience of previous sessions was, by far, the most important influence on my conduct of succeeding interviews. My growing experience resulted in a set of internalized responses and reactions which were more like reflexes than rationalized strategies. The very first group interview at Hanford is a case in point. As I mentioned earlier, participants had been initially brought together by Hanford officials for a program of lectures on the early history of the Project. The group interview also turned out to be, largely, another series of lectures. The responsibility lies in my own approach to the interview. The interview began by my turning to the first person at the table and inviting him to introduce himself and explain the conditions under which he came to Hanford and to describe the work that he had done on the Project. His rendition of that history went on for some fifteen minutes. In my mind, an eternity had passed. Doggedly, I turned to the second person, and then the third and on and on, until all five participants had spoken their piece. It was not an auspicious beginning to the project. The producer, James Hyder, and I reviewed that tape immediately after the session and mapped out a strategy in which the session would open and be moved along by a series of substantive questions or statements directed at the group rather than at any individual. It is a strategy from which I never departed.

A technique which proved to be very effective in group interviews was the use of historical photographs. During the war, photographic programs documented construction and operations at Hanford, Oak Ridge, and Los Alamos. I used some of these photographs during interview sessions, which provoked excellent exchanges on aspects of living and working at the site that might not otherwise have surfaced.

Because I was preoccupied with encouraging spontaneity between the participants, and because, with regard to senior scientists and administrators, I was determined to get behind a set of standard accounts which had naturally grown up over the years in response to many individual interviews, I never provided participants with a set of documents which identified the likely topics of discussion at the forthcoming interview. At various times I tried to introduce archival materials. The result was not good and led me to the conclusion that save for the simplest of issues, such material should not be introduced unless participants had been given a chance to review and digest the contents beforehand.

I distributed a number of original documents at one session. After several minutes the table was covered with paper. The participants were totally distracted from the flow of the interview and neither they nor I could cope with trying to assimilate the contents of the documents and at the same time carry on a coherent conversation. I did not make that mistake again.

In another session, I introduced only one document in the hopes of provoking a kind of discussion which the scientists and engineers often tried to avoid in other circumstances. The document was a memo from Captain William Parsons, head of the Ordinance Division at Los Alamos, to J. Robert Oppenheimer, director of the laboratory, written the day after the Trinity test. Parsons noted that they had all been surprised by the amount of visible radiation which had been emitted. In fact, for individuals close enough, looking directly at the explosion would have caused blindness. Parsons suggested that the effectiveness of the first bomb dropped on the Japanese might be increased if, just before the bomb were detonated, a device containing a loud whistle were to be dropped so that the Japanese would look up. I introduced this memo as a way of providing a context for a discussion of the effects of wartime fervor on the mindset of people who in other circumstances might well have cringed from such grisly thoughts.

The response to the document was very revealing. The participants read this rather brief memo in [what I interpret to be] stunned silence. Finally, one of them remarked that had they not been confronted with the document itself, they would have denied that it existed on the grounds that Parsons could never have written such a thing. While this remark in itself is revealing of the problem I was trying to address by using this technique, I was not able to engage them in a discussion of the issue I was raising. The lesson of this experience to me is that there is no substitute for preparing participants by providing them beforehand with documents central to the issues to be discussed. Other techniques must be developed for forcing those being interviewed out of protective shells of the sort we all rely on in thinking about the past.

DOCUMENTATION OF SITES AND ARTIFACTS

Production Values

In group interview sessions the issue of production values was limited essentially to the technical questions surrounding the need for broadcast quality. When, however, our primary goal was site and artifact documentation, there was a tendency on the part of the local contract production crew to organize "the shoot" as if the end product was to be a production rather than an archival product. This was especially true of the first three site visits where we contracted the services of local professional video studios who had not been part of the overall planning of the project and who did not have a previously established relationship with the producer.

Two examples stand out. At the very first session, which was to be devoted to documenting the first operational nuclear reactor at Hanford, I was asked to give a dramatic introduction to the session while standing in front of the face of the reactor. There was no written script and I had not given any thought at all to doing such a formal introduction. We did at least two "takes," neither of which was satisfactory from a dramatic point of view and had I not simply insisted that we move on, we would probably still be standing in front of the face of the reactor trying to get a smooth, dramatic introduction to the interviews.

The second example occurred at Oak Ridge near the beginning of our "tour" of the K-25 gaseous diffusion plant. We had just finished an interview inside one of the diffusion plant cells and were going to move next to a position outside the cells to discuss the operation of the main diffusion pumps. The director insisted that we begin the interview as if it were a scene, so that the editor of a production could edit between these two segments of the interview in such a way as to give the illusion that we had passed smoothly from one space in the building to the other. Nothing could have been further from the truth. Though the effort was totally out of character with the spirit with which I approached the series, to have interrupted would have been more trouble than it was worth and might well have destroyed what had developed as a very good relationship between those of us who had come from Washington and the local film crew. Later, quietly over lunch, James Hyder and I discussed the matter with the local director who, as it turned out, was quite responsive to our concerns.

This is not to suggest that rehearsals were not important. At both Hanford and Oak Ridge, camera angles in relation to my movements and the movements of those being interviewed were planned in order to make sure that the artifacts being discussed could be seen clearly. This was not always easy in tight spaces containing complex arrangements of equipment, for example in the reactor control room at Hanford or in the gaseous diffusion plant at Oak Ridge. After a while, neither I nor the local crews had any trouble interrupting the interview and beginning again. In the spirit of the archival nature of the project, all such false starts are part of the record.

The Eye of the Camera

In approaching the problems of site and artifact documentation the scholar has to be sensitive to his or her shortcomings with regard to what is visually possible. As someone who has had considerable experience in the field of still photography, I long ago learned that a person can learn to intuitively grasp the visual potentials of a scene. But I have also learned that the inner eye of the still photographer is not the same as the inner eye of the film and video photographer. His or her inner eye is sensitive to what the motion of the camera will reveal in a way to which the casual maker of home videos is blind. As we watch the unfolding of a film or video drama, we are not any more conscious of the emotive effects of the movement of the camera than we are of the role that focal length plays in the subtext of a still photograph.

This was demonstrated to me many times during the course of this series but most dramatically at the gaseous diffusion plant at Oak Ridge. When it was built in 1943–1944, this four story U-shaped building was the largest factory in the world. Each leg of the U is about 300 feet wide and a half mile long. Verbal descriptions cannot transmit to most people exactly what those figures mean.

Most of the machinery contained in the building was controlled by instruments located on the third floor. Because of the way the instrumentation is arranged, one can see from one end to the other. When we toured the floor to plan our set up, the local director, Robert Hasentufel, was awestruck at the possibilities his inner eye revealed to him. During the lunch break, he outlined for me a sequence in which we would place the camera in a motorized cart of some sort and just let it document a trip from one end of the building to the other. His perspective was a dramatic one and I immediately saw in his proposal a unique opportunity for documenting the gargantuan scale of this building.

I did not think Oak Ridge authorities would acquiesce to our request, especially since it required the use of a crane and the removal of an outer wall panel of the building as the only way to hoist a golf cart to the third floor. But in fact, largely due to Hasentufel's enthusiasm for the idea and his experience in dealing with such situations, official permission was granted and the footage was shot. This is only the most dramatic example of the many times useful suggestions of this sort came forward from the local production staff.

CONCLUSIONS

My initial hypothesis was that videohistory was a viable medium for two kinds of situations, group interviews and the documentation of sites and artifacts. My experience with the Manhattan Project series in the Smithsonian's Videohistory Program amply demonstrated that hypothesis. But we cannot ignore that, compared to the one-on-one audiotaped interview, videohistory interviews are expensive. Nor can one overlook the added administrative burden associated with group interviews. Getting everyone to the same place at the same time is not easy.

Is the added administrative burden and the extra expense worth it? That depends on the usefulness of the material. Transcripts of the audio sound tracks have exactly the same utility that any archival interview material has. The comparable use of the visual material will increase, I suspect, with passing time, especially as the individuals involved pass away and the sites and artifacts disappear. But if nothing else, we leave a visual record of many locations which were never before filmed. It is a shame that no one thought of doing the equivalent for the Los Alamos, New Mexico, site at the end of the war. It was all razed and the laboratory was relocated to a new site.

Lessons concerning the mechanics of the medium aside, there are insights concerning the Manhattan Project itself throughout these tapes. Most histories of the Manhattan Project focus on the scientific and technical challenges and achievements or on the effects of the development and use of the bomb on international relations. But one of the most striking impressions I received as I traveled around the country doing these interviews was the physical scale of the project. The Manhattan Project which is portrayed on these tapes is a very large and far flung industrial organization which employed about 100,000 people to build the factories and laboratories and over another 100,000 people to run the completed plants.

In spite of the scale and in spite of the large number of participants, virtually no one outside of Los Alamos who was not in top management had any idea of what it was that all this activity was directed toward. Participants who were not party to the secret testified again and again to the fact that much of the success of General Groves's compartmentalization program at the industrial level rested on a strategy of individual and group terror.

Another clear impression these tapes convey is the degree to which the Project became an end in itself. Even those Los Alamos scientists who later recanted, that is, who publicly regretted and denounced their enthusiastic involvement in building the bomb, can be seen in these tapes once again sucked into what J. Robert Oppenheimer described as "the sweet idea" as they recount with pride and enthusiasm their role in completing "the gadget."

In planning each session, I purposely excluded post-production use as a criterion. That seemed to me at odds with the notion of building a body of archival documents. Segments of this material should be useful in future productions in much the same way that other visual material not originally intended for production has always been used. Whether that proves to be the case or not, only time will tell.[2]

NOTES

1. There are two types of secondary literature I have in mind. The first are histories of the project such as the official AEC History, Richard G. Hewlett and Oscar E. Anderson, Jr. *The New World 1939/1946* (University Park: Pennsylvania State University Press, 1962), or the Pulitzer Prize winning book by Richard Rhodes, *The Making of the Atomic Bomb* (New York: Simon and Schuster, 1968). In these histories I was struck by the fact that they almost all portray the decision to commence an all out program to build the bomb as a consensual one which resulted from the progressive accumulation of technical knowledge and experience. I did not see this consensus in the primary documents. See Stanley Goldberg, "Inventing a Climate of Opinion: Vannevar Bush and the Decision to Build the Bomb," *Isis*, in press.

   The second type of literature are reminiscences of scientists who worked on the project and/or biographies of those scientists in which the administrative head of the project, Leslie R. Groves is characterized as a person of little understanding of the nature of the project, and as someone whose actions constantly interfered with the work of the scientists and engineers. See, for example, Nuel Pharr Davis, *Lawrence and Oppenheimer* (New York: Simon and Schuster, 1968). The documentary evidence suggested to me a person who was always in charge and who knew precisely what was going on. See Stanley Goldberg, *The Private Wars of Leslie R. Groves*, in preparation.

2. I am indebted to the Smithsonian Videohistory Program for their encouragement and generous support, especially David DeVorkin, Terri Schorzman, and Phillip Seitz. I thank my video producers James Hyder and Selma Thomas. Others who were particularly helpful were Andrew Szanton, who served as my research assistant for the Hanford, Oak

Ridge, and Cambridge sessions; Jay Haney, head of the Science Center at Hanford; and Ed Aebescher, head of the public relations office of the prime contractor for Oak Ridge. Deborah Reid transcribed the audio portions of the tapes, and Joan Mathys and Alexander Magoun of the Smithsonian Videohistory office tracked down all of the data necessary to turn those transcriptions into useful documents and finding aids. Ultimately, the responsibility for the quality of the footage in this series, good or bad, is mine. I am grateful that I was given the freedom to make these videos where and how I saw fit. I learned a great deal in doing them, not just doing video but about the Manhattan Project itself. I hope others will too.

VIDEOHISTORY AT WALTHAM CLOCK COMPANY: AN ASSESSMENT
Carlene Stephens

Our videotaping at Waltham Clock Company, Waltham, Mass., began as an emergency recording session to document the operation of historically significant automatic watchmaking machinery threatened with extinction. Waltham's executives permitted us to videotape what they believed would be the final operations of their factory. The firm was finishing its final government contract, and no others were expected. The six hours of resulting video contain rich details about worker skills, automatic machinery, and hand tools for making and testing timepieces.

Our principal goal was to document in operation machines identical to ones the firm had made available to the Smithsonian thirty years earlier. Although the filming at Waltham came about under the urgent conditions of the imminent closing of the firm, our work there fitted into the pattern of continuing research that Smithsonian curators perform to document museum collections.

The resulting video is a raw resource, personal notes about what we saw and heard at Waltham. A more detailed assessment of this experience and some reflections on the utility of video as a research tool for the history of technology follow.

WHAT WE KNEW BEFORE THE VIDEOTAPING

Beginning in 1849, Waltham mechanicians pioneered the machines and techniques for the mass production of pocket watches.[1] In the middle of the nineteenth century, making large quantities of *anything* by machine was still a novelty. To produce watches by machine, the firm not only had to invent watchmaking machinery, but also had to reinvent the watch, to simplify a tiny complex mechanism with hundreds of interconnecting parts. Waltham's technical history offers one of the most illustrative chapters in the development of American manufacturing. Waltham employees, once as many as four thousand strong, made tens of millions of watches during the company's long history.

Waltham inspired numerous competitors, both in this country and abroad. With two of the best known of those competitors, Elgin and

Hamilton, Waltham dominated American watchmaking through the nineteenth and early twentieth centuries.[2] By the late 1940s, though, the firm faced major reorganization tasks after war work, sold the name Waltham Watch Company, and reappeared under several new names with several new products. Most recently the firm was called Waltham Clock Company.

At the time of our first visit to Waltham in the spring of 1989, the firm was surviving largely on U.S. government military contracts and running the last mechanical watchmaking operation in this country. Waltham employees were making fifteen models of aircraft "clocks" for government contracts. The heart of each "clock" was actually an eight-day, very large mechanical watch movement, originally designed during World War II and now destined for otherwise extremely high-technology aircraft cockpits. Government specifications, according to Waltham executives, require this mechanical timekeeper in the cockpit to back up electronic timekeeping and navigation instruments.

The company invited us to film at a particularly difficult time in its long history. The very week we were filming, the firm notified its forty-four employees that most of them were finishing their very last government contract, and, as of the next week, they would be laid off.[3] The factory machinery was likely to be scrapped or sold. Even at the time of our filming, the firm had reduced the scale of its operation to a minimum, and rows of unattended, idle machinery filled most of the work rooms. In much the same way as the Historic American Buildings Survey or the Historic American Engineering Record is often one step ahead of the wrecker's ball in recording an important structure or site, we were there with our video crew to document as much as we could before it disappeared.

Because of Waltham's central role in the history of American manufacturing, the timekeeping collections at the National Museum of American History already contained a large, representative collection of Waltham-related materials. These included not only watches and watch movements, but also trade literature, illustrations of the factory building, engineering drawings of special-purpose machinery for making watches, and even the machinery itself. The Waltham machines had sat silent and unmoving in our exhibitions and storage areas since they came to the museum.[4] Although we had a rough idea of their various functions, we wanted to see in detail how they worked. We intended the video to complement the written and still-picture documentation we already possessed.

Many of us who study the history of technology have long recognized the utility of films produced by others as sources of information of this sort. Most of us need no convincing that the moving picture captures technical process and worker skill better than any other medium. Collections of industrial films are in many accessible repositories, and we make use of them regularly. We have located no other moving images of Waltham operations, though, so the prospect of producing our own footage was appealing because we hoped to get a detailed look at the machines.

A second use for the projected video occurred to us. Some of the watchmaking machinery will, in the next few years, go on display in a new timekeeping exhibition at the National Museum of American History. For a human operator to demonstrate machinery in an exhibition on a regular basis has never been feasible in our museum, but to demonstrate machines with a continuously running video is. We prepared the Waltham video to bring the machines to life, not only for scholarly purposes, but also for a wider audience.

WHAT WE FILMED AND WHAT WE LEARNED

We had to make some choices about what to shoot at Waltham. Some decisions were influenced by the limitations of the Smithsonian Videohistory Project budget, others were ruled by the site. Our video work at Waltham came late in the budgetary life of the project, and we had funds sufficient to spend only three days there. Even if our budget had been considerably larger, we still could not have filmed every step of manufacture. By the time we arrived, the factory was no longer performing all the steps. Even if they had been, the hundreds of operations to complete a "clock" would have taken weeks to film. Cooperative and flexible, the firm executives placed only one restriction on us. They asked us not to shoot the rows and rows of idle machines that filled the factory, because they feared the appearance might leave outsiders with a negative view of the place. Since our goal was to shoot operating machines, this restriction did not interfere with the content of our video.

We focused particular attention on operations relating to making and assembling the watch balance. The balance is an oscillating ring in the watch movement, equivalent in function to the pendulum in a clock, that controls the going rate of the watch. The balance consists of a metallic ring—often brass on the outside joined to steel on the inside—fitted with tiny adjustable screws and a coiled hairspring. The form and positioning of the various parts of the balance in relationship to each other critically influence the accuracy of the watch.

A Vander Woerds screw-making machine at the Waltham Clock Company was designed in 1870. [SVP photo, Terri A. Schorzman].

In a workroom dedicated to "Secondary Operations," the foreman, Joseph "Chuck" Martin, described the making and functioning of the balance wheel and demonstrated tapping holes in the balance wheel to insert screws. In a room devoted exclusively to making mainsprings and balance springs, Richard Halstead and Eddie Pitts demonstrated wire working and heat treating. And in the machine shop we filmed an automatic screw machine, designed about 1871, as machinist Chuck Paradis explained how brass wire stock became a tiny screw, at most one eighth of an inch long. It was this machine in particular that we had come to shoot, since it is nearly identical to one in our collection at the Museum.[5]

We had the unexpected opportunity to videotape not only this screw machine but two successive generations of machines. After World War II Waltham retooled and purchased Swiss-made screw-cutting machines.[6] We recorded a demonstration of one of these. We also videotaped a two-year-old numerically controlled machine tool made in Japan, two of which the firm had acquired to do a variety of machining unrelated to watchmaking. With video we were able to record, for the first time ever, three generations of operating screw-cutting machinery under one roof and discussions with the machinists who used all of them. We have preserved important details about the relative merits of each machine and, more importantly, recorded the changing relationship of skilled machinist to machine.

Videotaping at Waltham challenged our assumptions about the nature of watchmaking. These assumptions arose from a familiarity with the literature on the history of American watches and watchmaking, which falls into two main categories. The first is descriptive, collector-oriented, and focused on the physical and technical characteristics of individual watches or the range of products from a particular company. This literature is voluminous. The second, smaller body of watch literature addresses the history of watchmaking. These published accounts of factory production focus on the technical evolution of making watches by machine.[7]

If we can gain insights for studying the past from our recent experience at Waltham—and I think we can with certain caveats—one insight emerging from the video project is that published histories of watch manufacturing in this country have allowed a preoccupation with the marvels of innovative automatic machinery to overshadow the extent to which watchmaking relied heavily on handwork. Machines, tended by operators, made the watch parts, but skilled workers assembled and repeatedly adjusted them. And managers of both workers and machines designed the flow of materials, work, and finished products.

The prominent role of handwork in watchmaking is especially apparent in our video at Waltham. We filmed extensively in Final Assembly where supervisor John Valmas narrated as other workers assembled the movement and cased it. We were able to record some of the minute subassembly work for the balance and balance spring.

We also had an extended conversation with Stan James, a Jamaican-born, British-trained engineer, who did the final timing and testing of the watches. I was surprised and delighted to find an elaborate testing procedure at Waltham, not unlike those I have read about for testing marine chronometers in the nineteenth century. The ninety minutes of video we did with Stan James is the only moving picture or oral history documentation I know of anywhere that describes timepiece testing. Although our recording of these testing methods is unique, the video we shot of our conversation with Stan James is not particularly process-oriented and therefore most closely resembles traditional oral history. Stan stood in his lab, held up various models of timekeepers, explained the variations among differing models, and told us in detail, with minimal demonstration, how he tested them for accuracy.

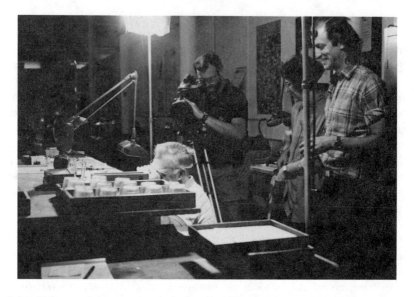

John Valmas, right, narrated the work of Edward Murphy while in Final Assembly. [SVP photo, Terri A. Schorzman].

Stan James explained the method of testing clocks to Stephens. [SVP photo, Terri A. Schorzman].

ASSESSING VIDEO AS A RESEARCH TOOL

Video's chief asset for conducting research in the history of modern technology is its unsurpassed ability to document and preserve process *dynamically*. It provides historians with a visual and audible record of people, tools, and machines *in motion*. Video also offers a means of sharing that information, in raw or edited form, with others.

An example from our Waltham shoot that demonstrates how this dynamism can be illuminating came during our conversations with John Valmos. As a curator of museum-held timekeeping collections, I had examined dozens of photographs of watchmakers at work and catalogued several watchmakers' benches, but I had never wondered precisely why the bench is built the way it is. During the videotaping session, the reasons became perfectly clear when I watched Valmos lean his elbows on the drawer of his own bench, use the work block at his chin level, and use a drawer below to catch imaginary materials he swept in pantomime from the bench top. I might have had this same insight just watching him at work without the video camera rolling, perhaps not. But the video shoot provided us with the opportunity to talk with Valmos and to watch him in action, an opportunity we would not have had otherwise. And the video now serves as a means for me to share this insight with others in a way very closely resembling the way I first experienced it.

Video's ability to record and preserve process dynamically provides historians of technology with an intriguing new research tool that promises to extend the possibilities of finding and using historical evidence in objects. This is a tantalizing prospect for those who embrace the techniques of material culture studies in the history of technology.

The object is the evidential center of material culture studies, an interdisciplinary field characterized by research that relies on objects for cultural information not necessarily available in written and statistical sources.[8] For most keepers of technical artifacts, whether or not they have adopted the discipline of material culture techniques, it is an article of faith that objects in their collections are useful historical documents. Collections of technical objects in major and minor museums have been around for a long time. But what Brooke Hindle pointed out nearly ten years ago is still true: the promise of object-based research in the history of technology has been only partially fulfilled.[9]

The promise goes unfulfilled, partly, because of the problematic nature of "reading" objects. Hindle noted that poor documentation and poor authentication of objects often makes them unreliable as historical evidence. Wrenched from their original contexts (put, for example, in a museum), objects offer only fragments of their original meaning or purpose. Faked or altered objects also obscure the historical record. And to the uninitiated, objects impart little information beyond physical characteristics like size, shape, and sometimes materials. "What can be gained from the visual examination of a technological specimen," Hindle wrote, "depends directly upon the knowledge and perceptiveness of the examiner."[10]

Understanding the history of technology through objects also is problematic because of the historian's impulse to reduce the experience of machines and process to words. But the fundamental problem of using words to write the history of technical thinking, machines, and techniques was described by Anthony Wallace in his book *Rockdale*:

> The work of the mechanician was, in large part, intellectual work. This was true in spite of the fact that he dealt with tangible objects and physical processes, not with symbols, and that some of what he did was done with dirty hands. The thinking of the mechanician in designing, building, and repairing tools and machinery had to be primarily visual and tactile, however, and this set it apart from those intellectual traditions that depended upon language, whether spoken or written. The product of the mechanician's thinking was a physical object, which virtually had to be seen to be understood; descriptions of machines, even in technical language, are notoriously ambiguous and extremely difficult to write, even with the aid of drawings and models.[11]

Wallace's description links the nonverbal aspects of machine design and the finished product. Our focus at Waltham was more on technical doing than technical thinking. We went to learn how machines worked in team with their human operators, not to learn what the machine designers had in mind when they constructed the machines, although we could have done both. But reliance on non-verbal information is still at the core of our inquiry.

Based on our experience at Waltham, it is clear to me that video can serve as the mediator many have sought between technical objects and verbal expressions of what they are and how they work. As a research tool, video gets at the fundamental non-verbal qualities of the history of technology (although, regrettably, we still cannot touch cold metal or smell hot oil through video). As an educational medium, video is a versatile and eloquent communicator, one that merges both spoken words and moving and still images to accommodate a variety of learning styles.

We learn from the video, as we do from watching in real time, not only what things look and sound like, but how they work. Sometimes we hear descriptions of techniques in the very words of the people who made or used them in their daily work, but more often we learn process in predominantly visual, nonverbal ways. No instruction booklet, written description, or static graphic explanation can substitute for this kind of firsthand experience, an experience that Brooke Hindle described as "feeling" the "three-dimensionality" of the machines and processes of technology.[12]

Making the video provides a direct experience with the machines or industrial process. Viewing the video afterward is not a direct experience but provides a means to analyze it later—to store, manipulate, and reproduce the information in the machines. After watching reruns of three generations of Waltham screw-cutting machines, I finally understand the sequence of steps to make a screw the size of a flea.

The nonverbal aspects of video have another side too. John Valmos was an exceptionally articulate guide to the processes in final assembly, but for the most part at Waltham we were documenting people at work who spoke little, knew little about the big picture of watchmaking outside their own set tasks, and did work that would otherwise go unrecorded. Video, like the study of objects, offers a kind of historical evidence that can balance some of the biases of verbal data, which in American history are, in Thomas Schelereth's words, "largely the literary record of a small group of mostly white, mostly upper or middle class, mostly male, mostly urban, and mostly Protestant cadre of writers."[13] The workforce at Waltham was, except for the managers, none of the above except urban. Noticeable were Asian and French-Canadian immigrants, who represent a changing ethnic mix in the town of Waltham.

Video also alleviates some of the trauma of wrenching objects for study out of their original context. Video shot on site, at factories and other workplaces, helps retain the context for the technical objects under scrutiny, which when removed to museums changes drastically. For example, despite Waltham's reluctance to let us film idle machines in the interest of avoiding the appearance that their operation is much reduced, showing the machines in multiples down the stretch of a factory room a block long conveys the messages of mass-production better than any other way.

The usefulness of video in recording objects makes the medium

especially useful for museums. Museums exist to acquire, study, preserve, and exhibit objects. Video can obviously aid cataloguing and
identifying, but much more. Video is one of several forces presently
moving museums beyond merely displaying and identifying objects.
Almost every new Smithsonian exhibition of moderate size, for example, contains explanatory video. Most are interviews with *people*
involved with making or using the objects on display.

Documenting museum collections in this way saves wear and tear
on the objects. Running them once for the filming eliminates the
need to run them continually for study or exhibition purposes. Our
tape from Waltham, for example, incorporated in a future exhibition,
will allow visitors to have the next best thing to a direct experience
with the machines. The video may be used, too, as a compromise to
prevent overuse of museum objects. Some machines might actually
operate a day a week, and then, at other times, visitors could have
access to videotape of them in action.

Video can also serve as a tool for note taking. But it surpasses note
taking with pad and pen not only for its ability to record nonverbal
content, but also in capturing detail. We couldn't possibly have taken
down every detail in written notes or remembered without some aide-
de-memoire all that transpired in the episodes we arranged. I see
new details everytime I review the tapes.

The tape's richness of detail is largely a function of the camera's
ability to look where the human observer cannot. This is especially
true in the case of watchmaking, where the scale of the individual
parts is so small and the fineness required to manipulate them during
manufacture, by hand or machine, is sometimes scarcely visible to
the unaided eye. The SI Videohistory Program administrator and
the Waltham producer, both experienced video viewers, told me that
the Waltham tapes are especially rich in detail because the cameraman was particularly talented.

Because the video is akin to personal notes, there is little question
in my mind that these tapes are subjective documents. They are a
reflection of not only the camera operator's and producer's skills, but
also the interviewer's skill (or lack of it) in asking questions, knowledge of the subject, and point of view.

The content of the tapes differs from written notes in an important
regard. We recorded on video for undetermined future uses, perhaps
an exhibition, and we recorded in such a way that a viewer without
my expertise in the subject could nevertheless understand what was
going on from the descriptions provided by the people speaking on
the tape. That is, even when I knew the answers to questions I asked

or knew very well what was about to happen, I nevertheless asked a series of questions to lay out the information in a logical sequence for a viewer who knew nothing about watchmaking. Rather than taking video notes on just new information, we sought within each segment a coherent, process-oriented storyline.

I learned two important lessons about technique that also significantly influenced the content of the videotapes from Waltham. Videohistory relies on the techniques of oral history, I now know. I had little experience with oral history, and Terri Schorzman, the project administrator, prompted me to learn to conduct the interviews in such a way as to let the interviewee do most of the talking. The result was less explanation from me about what we were seeing and more from the subject.

The other lesson involved watching the video monitor. Our initial setup did not give me a view of the monitor. But it quickly became apparent that the tiny scale at which we did most of our taping required me not only to conduct interviews, but also to watch the monitor to ensure that we got videotape of precisely what we were talking about in the interview, and conversely, talked about what we were seeing. Close attention to this kind of detail yielded the clear views we hoped for. This ability to watch the action is the chief advantage of videotape over film. One can see simultaneously on the monitor what the finished videotape will look like. With conventional film, only the person operating the camera has the view.[14]

Despite my lack of experience in alternative techniques, I did bring to the videotaping a specialized knowledge of watchmaking based on traditional written and graphic documentation. As curator in the Division of Engineering & Industry, I am responsible for maintaining and enhancing the timekeeping collections, and historical research on all aspects of timekeeping is routine business. In the context of the work of the division, I continue to see video as a *supplement* to other traditional and nontraditional methods of documentation for the full historical record, rather than a stand-alone product.

VIDEO OPPORTUNITIES

This supplementary role is far from limiting, though. The video experience we had at Waltham has suggested numerous new directions for future research.

Our video work at Waltham provided a microlevel view of watchmaking. If I were able to do more at Waltham, I would shoot more steps in the making of the aircraft clocks and focus next time especially on plate making and pinion and wheel cutting.[15] It would also be interesting to correlate the present factory layout, which is much smaller than at any other time in Waltham's one hundred and fifty year history, with earlier factory layout, with an eye to learning about the organization of work in the work space. Background information of this sort is available in other written documentation and the firm's collection of architectural drawings. It might also be interesting to extend the inquiry to pay attention to the design of the factory building and how its presence shaped Waltham's vernacular landscape. The complex, which transformed the Charles River in that area, once included—in addition to the enormous red brick factory building—a landscaped park, workers' housing, and other outbuildings. The factory once employed thousands of people, a significant proportion of the town's population, giving rise to its nickname, "Watch City." Waltham, as a historic industrial site, is ripe for exploration to answer questions about the past.

I would also like to do more video to document the state of contemporary watch- and clockmaking globally. The end of the 20th century is a crucial moment in the history of timekeeping technology. Electronic technology has eclipsed mechanical technology, except in the high-end specialty market. But the electronic technology is just barely over twenty years old. At sites in the United States, the Soviet Union, Switzerland, England, and Asia, I'd like to document the transition.

The kind of videodocumentation I'm proposing here is still not widely used. The technique is still in its infancy, and most historians have yet to become aware of the medium's capabilities and limitations. Not until the video camera is as ubiquitous as the xerox machine, I suspect, will it change the way historians do research.

NOTES

1. The name of the firm has changed numerous times since 1849. For simplicity, I'll use Waltham Watch Company, as the company became known in 1906.
2. For a business history of Waltham, see Charles W. Moore, *Timing a Century* (Cambridge, MA: Harvard University Press, 1945) and the chapter on Waltham Watch Company in Donald Hoke, *Ingenious Yankees: The Rise of the American System of Manufactures in the Private Sector* (New York: Columbia University Press, 1990).
3. The story has a happy ending for the firm. According to the government specifications, no imported parts are permitted for these aircraft clocks.

Customs officials raided the factory of the competitor who had been awarded the aircraft clock government contracts instead of Waltham and reportedly seized imported clock parts. When last I contacted the Waltham factory, the firm was in fact back in business, prospering, and branching out into other kinds of instrumentation.

4. The National Museum of American History has a twenty-nine piece collection of historical Waltham-related watchmaking machinery, gauges, and tools (museum accession nos. 225,117 and 260,024).

5. Automatic screw machine, Cat. No. 316,564, Acc. No. 225,117.

6. Vincent P. Carosso, "The Waltham Watch Company: A Case History," *Bulletin of the Business Historical Society* 23(December 1949): 179.

7. For watch articles, for example, see especially the *National Association of Watch and Clock Collectors Bulletin*. For a recent history of nineteenth century Waltham operations, see Donald Hoke, *Ingenious Yankees*.

8. Thomas J. Schlereth, "Material Culture and Cultural Research," *Material Culture: A Research Guide*, ed. by Thomas J. Schlereth (Lawrence, Kansas: University Press of Kansas, 1985), pp. 11–12.

9. Brooke Hindle, "Presidential Address: Technology Through the 3-D Time Warp," *Technology and Culture* 24(1983):452.

10. Brooke Hindle, *Technology in Early America: Needs and Opportunities* (Chapel Hill: University of North Carolina Press, 1966), p. 11.

11. Anthony Wallace, *Rockdale* (New York: W. W. Norton, 1978), p. 237.

12. Brooke Hindle, "Presidential Address," p. 456. Hindle wrote: "The historian of technology has to get inside the machines and processes of which he writes. He must feel their three-dimensionality. After he has established this acquaintance, he may be able to comprehend related devices through drawings and photographs or even words alone—but not without preliminary direct communication. Preserved artifacts, then provide an essential sense of technology, altogether aside from the specific factual information they may carry."

13. Thomas J. Schlereth, "Material Culture and Cultural Research," p. 12.

14. Thanks to Jack White for this insight, who drew on his experience with the filming of the NMAH-sponsored running of the John Bull, the world's oldest operable locomotive, in 1982. The event was filmed, not videotaped.

15. The plates of the watch movement are the platforms on which the wheels, pinions, and escapement are attached. The nature of the work performed on them during manufacturing is chiefly turning. The wheels and pinions are the gears of the timepiece. Pinions have fewer than sixteen teeth, wheels have sixteen or more. Making wheels and pinions requires cutting the desired number of teeth to a specified pitch.

ROBOT VIDEOHISTORY
Steven Lubar

Can videohistory provide appropriate tools for recording and understanding current science and technology? Contemporary science and technology is already the focus of the cameras of news reporters and science journalists, and, increasingly, the cameras of the scientists and engineers themselves. What can the historian who uses video bring to what seems a surfeit of videotape? Can the long-term view and background knowledge of the historian, mixed with the unsurpassed documentary value of the video recorder, provide a valuable new perspective on contemporary science and technology?

I believe that it can. I base this belief, and most of the discussion in this paper, on my experience working on the videohistory of walking robots. As part of this project, done (for the most part) under the auspices of the Smithsonian Videohistory Project, I videotaped four robotics projects:

- a student project to design and build a walking robot at the University of Maryland, in College Park, Maryland;
- a pick-and-place robot with vision, at the National Institute of Standards and Technology (NIST), in Gaithersburg, Maryland;
- a walking robot at Odetics, Inc., in Anaheim, California; and
- a robot designed for walking on Mars, at Carnegie Mellon University, in Pittsburgh, Pennsylvania.

I'll first describe these projects briefly, and then consider what I learned about the nature of contemporary robotics engineering—and, more generally, the workings of contemporary technology—by undertaking this project. Finally, I'll try to generalize about the value of contemporary videohistory, and give some hints for those who would like to use it for recording contemporary technology.

My first experience with videohistory was at the University of Maryland. There, I studied an undergraduate student project team that built a walking robot, Maryland's entry in a nationwide student competition. This was the third year of the competition, and the third entry from Maryland. The robot was designed as a cooperative project of a class of about twenty students, with the aid and direction of three faculty members of the Department of Mechanical Engineering and the Systems Research Center at the university. The first semester of the class was devoted to designing the robot, the second semester to its construction (and, as it turned out, extensive redesign). I sat in on most of the second semester classes, directed the videotaping of one class session and the laboratory session in which the robot walked for the first time, and reviewed, via videotape, the national competition.[1]

Next, I put my new videohistory skills to use on a small project at the National Institute of Standards and Technology. I had been offered, for the collection of robots at the Smithsonian, a pick-and-place robot that had been used in NIST tests of vision and robotic hand-eye coordination. I was happy to accept the robot, but realized that conventional documentation—sketching out the circuits, photographing each of the electronic boards, and so on—would be expensive and time-consuming. Moreover, it would not be particularly useful for understanding how the thing worked, or its history. Instead, I arranged for the chief engineer and chief scientist who had worked on the robot to spend some time with the robot in front of a video camera. They described each part, explained how the robot worked, why it was built the way it was, and how it had changed over time. They also discussed the experiments that had been done with the robot. The robot no longer worked; the only action shots were of the engineer moving the robot through its routines.

Odetics, the firm that invented the ODEX, a walking robot it calls a "functionoid," was my third videohistory site. I learned of the robotics work at Odetics through a donation to the Smithsonian; we were offered ODEX I in 1986. This videohistory session had three goals: to document the ODEX we had in our collections, to document the work of the engineers designing its successor, ODEX III, and to document Odetics' corporate culture. Odetics still had the twin of the Smithsonian's ODEX I (slightly updated and renamed ODEX II), which made the first part of the project possible. ODEX III was in the debugging stages. It was mechanically complete and its control software was being fine-tuned. Both its design and the process by which it was designed were dramatically different from the Smithsonian ODEX. We hoped that the comparison of the two projects would be revealing, and, indeed, it was. The third part of the project was focused on Odetics as a corporation. It's famous as a "good place to work," a laid-back southern California company that does research and development on the cutting edge of technology and yet manages to turn a profit.

My fourth, and, to date, final robotics videohistory session was at Carnegie Mellon University's Robotics Institute. Carnegie Mellon is a leader in robotics, and the project I studied there, a twenty-five foot tall robot designed to roam the surface of Mars, is "state of the art" work. The robot, called Ambler (a forced acronym for Autonomous MoBiLe Exploration Robot) makes use of a complicated network of inter-communicating computers, and is being designed in a cooperative manner by a team of students, staff, and faculty. Ambler is funded by the National Aeronautics and Space Administration, and is supposed to be completed in 1992.[2]

To all of these projects I brought the mindset and interests of a historian of technology. I wanted to answer two fundamental questions, to my mind the two questions essential to understanding the history of technology, or, for that matter, current technology:

- What shapes the development of technology? Convinced, like most historians of technology, that more than "mere" technical reasons are involved, I wanted to discover and document the forces that shaped the robots: the previous experience and beliefs of the designers, the knowledge and skill base upon which they drew, the institutional surroundings which helped or hindered them, the markets and audiences they hoped to please.
- What is the nature of engineering thought? Or, as a recent book put it, "What do engineers know, and how do they know it?"[3] What sort of knowledge is the base for making engineering decisions, and how are those decisions made?

These questions are usually applied to understanding historic technologies, but they have equal utility for guiding research into present-day technology. Moreover, it seems that there might be some shortcuts in their application to contemporary technology. It would be nice to be able to ask James Watt what he *really* thought when he first took a good look at the steam from his tea kettle, or to videotape Thomas Edison as he invented the lightbulb. On the other hand, it's possible for the historian to lose historical perspective in trying to answer contemporary questions.

My experiment in videohistory aimed at more than just finding new documentary techniques. More fundamentally, I wanted to know if the video examination of current engineering could tell us something about the nature of engineering thought and activity. Were the topics of interest in the history of technology applicable to contemporary technology? And how might the techniques of videohistory help in answering them, and in refining their application to past technologies? Those were the questions I had in mind when I set out to document current robotics research with camera and camera crew.

My use of videohistory was more a research tool than a tool for documentation. The two purposes, however, are closely linked. To *document* a process or an object fully one must *understand* and *investigate*. Understanding can be a by-product of careful, detailed, documentation just as understanding can result in a better documentation. Still, there is a tension between documentation and investigation. That tension, reflected in the contrast between details and big picture, shapes the process of videohistory.

Videohistory forces attention to the details. After all, what the camera picks up *are* the details: expressions, actions, words. The videohistorian is overwhelmed by a sea of details, and must select among them and make sense of them. The emphasis on details also derives from the social nature of the videohistory project. In order for the others on the videohistory team (the producer, the camera crew, and the sound crew) to know what's going to happen next, the videohistorian must think ahead and indicate what he or she is after: what shots are important, who should be miked, what's going to happen. Every detail must be planned.

For the same reasons, the subjects of videohistory must be given advance notice about the direction of the interview. This has its good and bad sides. Unlike video journalism, where the surprise question can elicit revelations, and unlike *cinema verité*, where long hours of living with the filmmaker and camera crew make it possible for the subject to forget their presence, videohistory is not good for discovering what the subject doesn't want to tell you. On the other hand, it's good at eliciting from participants things that they don't even know they know, but which are critical to their work.

Videohistory is thus useful, paradoxically, for getting not at details, not at individual stories, but rather at the deep structure of people's work. It can reveal how people think, not just what they do. In part because of the complexity of the setups required, their cost and the time involved, videohistory is a blunt-edged investigative tool. It is best not at finding specific answers to individual questions, but rather at finding general answers to bigger questions. Videohistory is a net, not a spear. It records details but captures the bigger picture.

As a documentary tool, the basic techniques of videohistory are those of oral history, but oral history supplemented by a large visual component. My first technique was the straightforward interview. One on one, or in small groups, I gathered the important parties together around a table, or, better, around the robot they had designed, and asked them to tell me their story. If the object or objects they were talking about were present, then almost without fail my interviewees held them up to the camera and explained how they worked in motions as well as words. Engineers, historians of technology suggest, think three-dimensionally, in terms of action and motion. They imagine objects sweeping through through space and time. They visualize the three-dimensional shapes and actions of a design even before it's created or set in motion. My interviews bore these assumptions out. An audio tape recorder would have missed the hand waving and the movements of mechanism—so often exactly what I was after. Video was essential to capturing their actions, and, through their actions, their ideas.

The use of objects in motion as a key element of description varied between projects, though its presence was apparent in each of them. And indeed, there seems to be some consistency in the way mechanical engineers, at least, thought about their work in the projects I examined. This was most obvious in the University of Maryland project. Students who were still in the process of learning to be engineers were clearly not as good as their professors at visualizing design, or the ways in which the robot would move in space, or in which its parts would move with respect to one another. The video captured the students' inexperience both in their lack of explanatory ability—they didn't have the vocabulary, either verbal or physical—and in the design, construction, and performance of their robot.

There were also interesting variations in the ways in which different types of engineers graphically communicated their work. Mechanical engineers would use models to explain their designs, or, lacking a model, wave their hands to indicate motion. The closer the engineer was to the production process the more likely he was to use these techniques. Software engineers, on the other hand, were better at verbal explanations, and tended to resort to paper explanations, drawing diagrams as soon as a concept became difficult to explain in words. Video captures these techniques better than any written description could.

Visual, three-dimensional thinking is also a major part of the process of building machines. Student machinists at the University of Maryland had a hard time predicting the outcome of a manufacturing decision, while the manufacturing engineer at Odetics had an astonishing ability to do so, and, in a surprisingly common linkage, an excellent ability to explain it. Time and time again I was pleasantly surprised by the skill with which engineers could explain their work.

These explanations were clearly helped by the presence of the artifact. The artifact acts as an *aide-mémoire*. Indeed, the artifact seemed to encapsulate, for the engineer, every decision that went into its con-

Armen Silvaslian explained the design and testing of the telescoped legs for ODEX III. [SVP photo, Terri A. Schorzman].

struction. Perhaps the most directly valuable result of videotaping with the object on hand, was the clarity of reminiscence and explanation that the object seemed to allow. The presence of the camera encouraged, rather than discouraged, this sort of explanation. Engineers were glad to explain an object for the camera, even when they were hesitant to participate in a simple interview. The evocative effect of the artifact was true even in the case where engineers were discussing work they had done several years ago, as at Odetics and NIST.

I increasingly emphasized the three-dimensional and moving aspects of the interview as time went on, spending more time on "how things worked." Descriptive videohistory drove out investigative videohistory. My interviewees were most comfortable with this sort of description—more so than trying to describe how or why an invention came about. They were used to this process as part of their everyday work as engineers. And the descriptions were sometimes astonishing. The language they used, both verbal and physical, was well developed. Shy, untalkative interviewees became eloquent as long as they focused on the objects themselves. Occasionally they spoke as though from the perspective of the robot, or mechanism, or computer. More often, they spoke of their machines as an "other,"

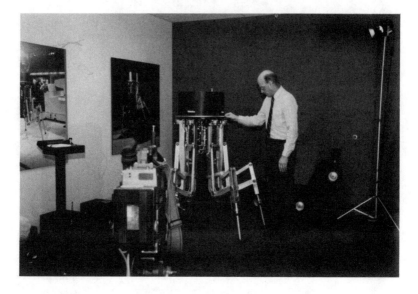

Steve Bartholet demonstrated the operation of ODEX I and often referred to machine parts as "his elbow. . . . " [SVP photo, Terri A. Schorzman].

almost as a person. The machine was often "he," not "it." Sometimes, each part of the machine was given its own voice: *he* does this, and then *he*—another part of the machine—does this. Video was ideal for capturing the interaction of technologists with their technologies.

In my attempt to capture ongoing engineering work, I tried a second technique: video as documentary tool. Here, videohistory borrows techniques from video journalism and, to a lesser degree, *cinema verité*. I used two basic techniques. I set up in a room where the engineer or engineers were at work, and watched and recorded their work, occasionally asking them to explain what they were doing, and why. This was expensive in terms of both time and tape, and was more intrusive than I would have liked. However, it was surprisingly effective at capturing something of the process of engineering, especially software engineering.

Software engineering—the writing of computer programs—clearly does not lend itself to the use of visual models in the same way that mechanical engineering does, and so it was not clear at the start that we would be able to capture the essence of software engineering on tape. After all, programmers undertake an activity which looks, from a distance, like word processing. It's impossible for the camera to focus closely enough, or for a long enough time, on the screen to capture the details of programming, or to figure out the whys and wherefores. Not only that, but the results of most software engineering are invisible anyway, or if visible, not very exciting.

Indeed, one of the reasons I picked robotics for this study was because I hoped that the combination of mechanical and software engineering that is essential for a robot would make visible the results of programming. When a key is hit, not only do the words on the screen change; the robot moves, too. This visualization of software worked better than I expected, and I was able to capture on tape debugging, an essential element of software engineering.

Debugging is the art of fixing computer programs. In some ways, it's analogous to rewriting. Just as the first draft of an article is not argued as carefully or stated as felicitously as it might be, software rarely works correctly the first time. Robot software is very complicated, and so there is often no obvious reason for the robot's actions; the robot often seem alive. Their actions become a puzzle that the engineer must figure out. I documented the art of debugging—the solving of these puzzles—by watching the robot's activities, and then recording the conversation within the group of engineers as they described their perception of the failure and tried to solve it.

The best debugging sessions we recorded were those at Odetics and Carnegie Mellon. At the time we visited Odetics, ODEX III, mechanically complete, was mounted on a test stand so that it could "walk in place." A sophisticated test system had been designed to debug the walking algorithms, that is, to make sure it didn't trip over its own six feet. The engineer running the test made minor changes in timing and length of stride as he watched the robot go through its walking motions. The camera focused on the robot, the test screen and the engineer as he worked, recording his running monologue about the problems he saw, and why and how he was making changes. It captured, in visual terms, his thought processes.

This process worked even better when there were several people working together on one problem. Turning the videohistory camera on the debugging of the Ambler at Carnegie Mellon allowed us to capture another aspect of software engineering. The Ambler, like many robots, uses several computers, all communicating with one another. The project management followed the same lines, with each computer having its own team of programmers. Debugging took place in a small room. Each programmer sat in front of his own workstation. He'd call out to the others what he was doing, what his computer was doing, and what messages it was sending to or receiv-

Steve Corley made changes in the walking motion of ODEX III by debugging the algorithms. [SVP photo, Terri A. Schorzman].

ing from the other computers. The camera and microphones captured this, as well as the occasional motions of the robot just outside the control room.

Mechanical designs too need debugging, and the contrast of that activity with software debugging made for an interesting study. The students at the University of Maryland class were building and debugging their robot's mechanical design at the same time. (They knew that this was not good practice, but because of the peculiar nature of the class, and the time pressure of the competition, they had no choice.) They had a hard time predicting how it would work. Not until they saw the pieces of the robot assembled were they able to see what happened when it moved. They then had to redesign and try again.

In each of these cases, videohistory captured a general sense of the nature of the activity, not a detailed understanding of the problems being solved. Videohistory as a means of documenting ongoing work cannot capture the narrative of a particular solution—you're not there long enough for that, or at the right time—but it can capture overall patterns of activity very well indeed. The medium is especially good at revealing the patterns of intra-group communications.

Engineering, videohistory makes clear, is a social phenomenon; it's done in teams, and communication among engineers is as important as communications among the computers they're programming. While not particularly successful at capturing the management meetings—there's nothing there for it to focus on—videohistory reveals superbly how teams work and how management effects the way engineering is done. The camera captured, for example, the changes in management style at Odetics, from the "Skunk Works" whatever-it-takes-to-get-it-working style that produced Odex I in a matter of months to the careful, methodical style that was taking years to produce Odex III, but which would produce it in a way that was completely documented and easier for people other than its designers and builders to use. Videohistory captures engineering style very nicely.

While the videohistory sessions I undertook are hardly typical, even of work in engineering, it is possible to generalize from them and to give some hints for those who would like to try videohistory for recording contemporary technology.

One way to do this is to consider the differences between the footage taken for videohistory and the miles of videotape taken for other reasons, either by the engineers themselves or by TV crews for documentaries or news program. Unlike the TV camera, our camera focused not on events but on the theory, design, and process behind the events; not on the immediate but on the long view. I went with questions in mind, questions based on historical understanding of the work I would see. Even more important, in videohistory documentation, the videohistorian interferes in the scene: he or she pushes the situation, asks questions, brings up counterexamples, and demands explanation. You've got to make the subjects explain what they're doing, and why.

Our videotape is also different from the videotape the engineers take of their own work. We focus on different topics. Their video tends to focus on technical details of value to those who already know everything about the process. Ours steps back a bit, asking bigger questions of more general interest. Ours is indexed and archived; theirs piles up unexamined and inaccessible.

But perhaps the most critical difference is that videohistory is—or can be—more honest. Perhaps the greatest skill required of the videohistorian of engineering is the ability to determine what is real and what is fake. Every engineering project develops canned "demos" to show off to visitors, funders, and higher-ups in the laboratory. It's quite easy to make a machine look like it's working, or to make it work in special circumstances. After all, the casual visitor can only see the machine do what the engineers offer to show. Even in these limited cases, you can't see what's really going on inside the machine: is the robot simply reading a script, or is it really reacting to its environment the way it would have to if it were out in the real world? It's easy to fake robots. Indeed, the whole history of robotics is filled with examples of robots doing impressive feats at news conferences—robots which are never heard from again because they didn't really work in the real world.

Perhaps the real art of videohistory, and its true value, is in capturing the process of engineering and science with a greater verisimilitude than is otherwise possible. It doesn't capture *progress*—that's easy to do after the fact—but rather *process*, the day to day work of the laboratory. Good videohistory, by combining in-depth interviews, slices of daily laboratory life, and detailed explanations of artifacts in motion, allows the historian to understand how things happen: the sometimes dull but nonetheless important process of engineering work. That's an important catch in the videohistory net.

Because videohistory serves both research and documentary purposes, and because it focuses on the details but answers bigger, more general questions, videohistory projects can tend toward schizophrenia. Video is often better at style than substance, and so it's possible to lose sight of the goals of the project. Also, because it captures process, not progress, videohistory can seem incomplete; it seems to lack closure. But in these contradictions lies videohistory's strength. For by documenting selected details of a project, it becomes possible to understand it in the most general way. Videohistory is documentation and investigation rolled into one. Because of that, it can offer both a new and deeper understanding and a fuller and more accurate record.

NOTES

1. This class is described in more detail in S. Azarm, J. Chen, and L. W. Tsai, "Walking Robot: A Multidisciplinary Design Project for Undergraduate Students," *International Journal of Mechanical Engineering Education* 18 (April 1990).

2. For more on the Ambler, see J. Bares et al., "Ambler: An Autonomous Rover for Planetary Exploration," *IEEE Computer* (June 1989): 18–26.

3. Walter G. Vincenti, *What Engineers Know and How They Know It; Analytical Studies from Aeronautical History* (Baltimore: Johns Hopkins University Press, 1990).

PRESERVING A TOOL-BUILDING CULTURE: VIDEOHISTORY AND SCIENTIFIC ROCKETRY
David DeVorkin

INTRODUCTION

One of my many goals in exploring the origins of space research in the United States is to better appreciate how the space sciences emerged as a tool-building culture within the national security state.[1] Tool building in this case means the activity of creating devices to perform certain functions on rockets, like obtaining the ultraviolet or X-ray spectrum of the sun, or capturing a high-energy cosmic ray. A tool-building culture in this context, then, is one which is interested primarily in building things that are useful, somehow, to someone. Historian of technology Melvin Kransberg was once overheard saying, "The scientist aims to understand nature; the technologist aims to make useful things."[2] Thus pioneer American space scientists indeed acted as technologists. My interest then as I explored this aspect of the early space sciences was to recover those characteristics of the culture that best revealed its nature. Without question, this meant finding pioneers in the field and putting them in touch once again with the instruments they created to explore the earth's high atmosphere with captured German V-2 missiles. These were 14-meter high, 12,800-kilogram liquid-fueled rockets created at Peenemünde in the early 1940s and launched on civilian targets in Belgium and England in 1944 and 1945.[3]

After spending five years studying the origins of the space sciences, using traditional library and archival techniques as well as extensive oral histories, I found that most of my questions could be answered quite satisfactorily. Most definitely, traditional modes of inquiry yielded critically important information on who these people were, where they came from, who was interested enough in the work to support its enormous expense, and what was accomplished. But there was something lacking as I became familiar with these groups. Hints came from original notebooks, laboratory test reports, and personal correspondence that helped to reconstruct what was going on in the laboratory. Specifically, why did some groups succeed, and others fail? What were the attitudes of those involved toward the science they were doing, and the instruments they were building to perform scientific research? When these people walked into their laboratories, what did they see and do?

Two factors allowed me to explore these questions. First, vestiges of original space science laboratories still exist at the Naval Research Laboratory in Washington, D.C., along with their creators, and second, the Alfred P. Sloan Foundation made it possible to explore the use of video documentation in historical research.

Much of what is discussed in this paper is based upon a larger study employing primary documentation. I focus here on how the tool-building character of two NRL groups, one headed by Herbert Friedman, and the other by Charles Johnson, have been captured using oral and videohistory techniques.[4]

SCIENTIFIC ROCKETRY AS TOOL BUILDING

Traditional modes of inquiry taught me that these groups were highly focused on the technical problems involved with making their devices work in the realm of a rocket, and that none of them were previously part of the scientific disciplines traditionally interested in studying the upper atmosphere. Francis Johnson, one of the very few at the Naval Research Laboratory (NRL) who did have training in any form of atmospheric physics, made the latter point in oral history testimony:

> It seems to me that when the rockets came along, they provided a great opportunity which lots of different people saw ways to make use of. It didn't necessarily reflect a prior interest; in many cases it didn't. It was: "who had devices that could be flown?"

Johnson added:

> Maybe Friedman's work typifies that best of all. He was working with various types of counters, and he had the ability and the people who could assemble such things and put them in rockets. So the interest developed there because of the opportunity. It wasn't that he had been waiting a long time for an opportunity to make these measurements on the sun, or on astronomical objects.[5]

Herbert Friedman's training in solid state physics at The Johns Hopkins University, and his experience during the war developing reliable gas-filled photon counters, gave him an expertise valuable for a wide range of applications. A radiation detector he created during the war was reproduced several million-fold by the Navy's Bureau of Ships; and, as Friedman recalled during a recent videohistory session, this provided him with enormous equity after the war as he continued to apply his counter to a wide range of Navy applications.[6]

As a tool builder, Friedman was in possession of highly desirable technical knowledge, and the Navy was keenly interested in exploring all possible applications. When the V-2 rockets were first flying, Friedman was deeply involved in nuclear test detection and in dosimetry, but when the photographic efforts at capturing the ultraviolet and x-ray spectrum of the sun seemed to be stalled in the late 1940s, Friedman entered the rocketry field. As he recalled:

> There was no way the people who were using spectrographs and photographic film would ever get down to the X-ray region. So it was a very natural thing to think of using our counter techniques to do X-ray photometry at the first opportunity.[7]

By 1948, Friedman was adapting his counters for these spectral ranges, knowing that his technology was comparatively well in hand, even though to most outside academic physicists, photon counters were still a mystery.[8] Friedman's goal was to solve a recognized problem the Navy was keenly interested in: what was the specific cause for ionospheric disturbances that could affect long-range radio communications? This institutional goal, and his established expertise in detector technology, placed Friedman and his associates squarely where scientific rocketry intersected with solar and ionospheric physics.

REFLECTIONS ON
THE TOOL-BUILDING CULTURE

In a preliminary group videohistory in 1986, Friedman and charter members of his staff recalled their roles. Edward T. Byram, for instance, described his early responsibilities "helping to find good Geiger tubes that had stable characteristics, and proper sensitivity."[9] But what was surprising was that Byram, with many of his original detectors sitting on the table before him, never took his eyes off the tiny metal chambers, and when asked if his responsibility for finding the best ones was not at times frustrating, Byram quipped back:

> I was never frustrated. I enjoyed fighting them. It wasn't a frustration, it was a challenge. It was mind over Geiger tube. [Everyone laughs]

And when asked how he made sure that he had control over these cantankerous devices, he added: "By keeping at them. Just looking at them more or less constantly, checking them." At this point, the interview took a turn. Up to now, the pattern had been that I would ask a question, and get an answer. But Byram's quip caused others to reflect. A younger colleague, Robert Kreplin, then added:

> I think we started out with ten or twenty counters of the type that you wanted to fly one of, and [by] filling them, and then checking plateaus and thresholds as a function of time, day by day—pretty soon you picked out the ones that were leaking and tried to repair and refill those, and hope[d] by the time it came to install them, you had at least one left that was good.[10]

Byram was focused on the immediate needs of his detectors, whereas Robert Kreplin assumed both technical and administrative responsibilities. His perspective reflected his different role. Friedman, always the leader, defined the agenda for the group, and defended it to the outside world. All the others looked inward, and made the devices Friedman had designed operate in the realm of the rocket.

"WE JUST DID IT"

Friedman's group translated basic designs into prototypes and tested them for proper spectral response, quantum efficiency, and reliability. The sophisticated specialty shop facilities NRL maintained for glass, electronics, and mechanical fabrication were essential in their tool-building culture. Friedman could simply "sketch a tube to them. They would convert it into a design which could be put together."[11] As a result, his staff left few formal directions on how to build these new tubes, and very little documentation appears in surviving laboratory notebooks.

Well-defined responsibilities but flexible procedures defined Friedman's group structure. As an exploratory unit, it was not unlike others within NRL or, for example, at the University of Michigan's School of Engineering, where another rocketry team leader defined the exploratory agenda and assembled the expertise needed, mainly from his former students and assistants. Nelson Spencer knew that his Michigan boss, William Dow, wanted experimentation to be open-ended and collegial: "We never had any formal meetings or any formal organization or anything. We just did it."[12] This was the hallmark at NRL. When asked about blueprints for building vacuum chambers and experimenting with mass spectrometers, Charles Johnson, who was in an ionospheric physics group working parallel to Friedman, thought back over a life in the laboratory, and shrugged his shoulders:

> Well . . . in our work here, we often said, "You just do it." (Laughs) How you "just do it," we don't know, but we did it. We just had to do those things that were necessary. We took our ideas and put them down in sketches, on paper, [and] took them over to the design and drafting group, [where] they would put them up in nice drawings to build the equipment, [and] then the shop would make it for us. Other than that, we had to use our own ingenuity. You're faced with solving a problem; you do it.[13]

Johnson's world was a world of trial and error, looking for vacuum leaks and plugging them, and constantly searching out more efficient and powerful diagnostic tools.

"TESTING TESTING TESTING"

Charles Johnson, like Byram and Spencer, focused on instruments, poking and prodding them until they worked. During oral and video interviews, they all tended to forget about paperwork and bureaucracy. Their recollections were clouded by more recent experiences: the mountains of bureaucracy they encountered in the NASA years made the earlier paper pushing pale by comparison. After discussing their collective experience in NASA's Project Ranger, Talbot Chubb and Kreplin compared that experience with what went on before. After Chubb recalled that testing always had highest priority in the NRL shops, Kreplin added, as if reciting a dirge:

> The philosophy was just to test it and test it and test it until you were convinced that it was going to work.[14]

Charles Johnson felt that testing was his life. When asked how long it would take to make sure a mass spectrometer was working properly, he visibly sighed, again thinking back,

> A lot longer than it's taken me to tell you. (Laughs) It's weeks . . . to do a thing like this, we learn that you constantly keep testing, testing, testing, to see that everything stays just like you want it. The last test is when you fire it. And if we didn't take that attitude on it, we got into trouble.[15]

Byram's recollections and body language paralleled Johnson's; when an instrument they had built was in sight, their energies were directed toward it as if they were reliving the experiences they had bringing it to life. In the 1960s and 1970s, Byram and Kreplin had to work under a NASA policy that dictated destructive prototype testing, but did not include testing the flight instrument. This policy effectively removed them from complete control over what they flew on satellites. During a video session held in an NRL studio in 1987, where the group examined dozens of detectors and instrument components, Byram bristled with the memory of testing a gas chamber for their High Energy Astronomy Observatory (HEAO) detector array:

[DeVorkin places a burst HEAO gas chamber on the table]
BYRAM: The requirement of NASA was that this be tested at four times its rated pressure. We were going to operate at 500 psi, so the test had to go to 2000. It went to 2000 successfully, but they didn't stop the test; they ran it on up and blew it up.
DEVORKIN: Did they want to test it to destruction?
BYRAM: They wanted to, but if I had been there, they wouldn't have done it.
KREPLIN: That illustrates the difference between our tradition growing up in the space business, running our own shows, and then getting involved in the very large programs that NASA was running. The management levels get piled one on top of another, so that now working with the shuttle, we find that in a case like this, it's necessary to build four of these to get one, because the first has to be tested and burst; the second one has to be tested, cycled, and then burst; and the third one has to be cycled. Again, this is two levels greater than flight. And the fourth one you can use.
DEVORKIN: Assuming the fourth one has the characteristics of the first three.
KREPLIN: Well, that's assumed, I guess.
DEVORKIN: How comfortable do you feel in those types of assumptions, especially when you're dealing with contractors?
BYRAM: Well, I don't like the idea personally.
DEVORKIN: You do prefer to fly one that you test?
BYRAM: Right.[16]

The purpose of this video taping session was to have Byram and Kreplin reveal, through discussion and demonstration, just what was involved in designing and testing the dozens of different detector systems Friedman's group had created between 1950 and the 1970s. Having the detectors at hand created an immediacy that stimulated old feelings of frustration as much as they brought back many intricate design details that helped to rationalize why each detector looked

E. T. Byram, left, and Robert Kreplin, right, discuss the burst HEAO gas chamber during their interview with David DeVorkin. [Still image from SVP video footage, NRL video crew].

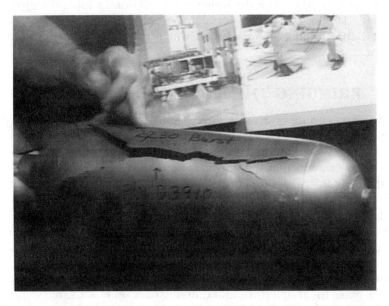

Closeup of the burst HEAO gas chamber. [Still image from SVP video footage, NRL video crew].

so different. Numerous technical procedures surfaced in these sessions, such as how to find optimum filter/gas combinations that would provide reliable and reproducible responses. Robert Kreplin described design problems while looking at some of the earliest detectors, designed to detect the Lyman alpha region, and recalled that many times the best filler gases, such as chlorine, would attack the silver that connected the filter to the tube walls. But when they turned to another type of quench agent, like nitric oxide, which could be ionized by Lyman alpha quite readily, they also found problems:

> Unfortunately, [nitric acid] reacted with, as I recall, the oxygen on the surface of the stainless steel. So it was necessary to build some detectors with oxygen-free copper. Also, because of . . . reactivity to [impurities] in the detector body, the stability of these tubes was not really very good. We had to do a lot of filling to get a few tubes, and their lifetimes were not all that great.[17]

Some of these complexities could be found in the old laboratory notes, but handling these artifacts and being asked structured questions about them brought back detailed recollections like these from Kreplin which helped to better understand the often disconnected commentary in the notebooks at NRL. The many compatibility problems also help to explain the wide variety of counter tubes the group created. Different filter/gas combinations often required different tube wall or sealing characteristics, which made each tube look different.

BRINGING THE LABORATORY TO LIFE

We examined detectors in a studio setting, which was only a partial step back to the laboratory, where the detectors were given life. Taking the next step, Talbot Chubb led us through a filling procedure at a filling station that was within one generation of those he used in the 1950s.[18] Even though the configuration of the new station had changed, the old design was so ingrained in his tactile memory that on many occasions, his hands reached for valves that had not existed for decades. After describing the detailed operational differences between the present design and the one they had used, Chubb demonstrated how the old system worked, and as he did so, he operated a system that survived only in his memory. Handling the intricate glass/Invar detectors, and thinking back on how the old station worked, Chubb talked about the old support structure he enjoyed at

NRL when they worked "with a lot of technician support . . . I worked very closely with the glass blower in developing the special ultraviolet-sensitive photon counters which we used and the soft X-ray counters."[19]

As Chubb handled the tiny glass detectors, he recalled how essential the NRL laboratory services had been to the program. He also remembered that many of the shop people "became design technicians, who actually designed the structures of our rockets and the housings that held our detectors."[20] It was very much a sense of team work that emerged as Chubb relived the delicate process of designing, building, and testing these devices in the old laboratory he had grown up in.[21]

Charles Johnson and his younger colleague Julian Holmes provided a revealing glimpse of what it meant to work as a team during another video session. While Johnson explained how their mass spectrometers worked, Holmes instinctively turned to a paper easel and drew a diagram of how ions passed through the front stages of their mass spectrometers to illustrate Johnson's words for the camera.[22] This was a helpful spontaneous action, although we may have anticipated it by making the easel available. Even so, it was clear from examining NRL laboratory notebooks and from other group interviews that spontaneous helpful action emerged from the clearly identified yet shared responsibilities of each group member. Holmes acted when something was needed; as he rushed to meet launch deadlines, each member of an instrument group sensed what was needed without being told what to do at each step.[23]

Johnson and Holmes felt strongly that within their immediate instrument group, everyone had to know a bit of everything to be adept at trouble shooting. As Holmes recalled, "I was brought up in the program where you did the job from beginning to end." Over time, each member did a little bit of everything:

> Everybody in our group was, to a great extent, a hands-on person . . . When you went into the field to fly it, you worked on all aspects of that operation . . . so that no one person was indispensable. But Charlie had to run the whole thing, so he was really indispensable.[24]

Teamwork translated into shared responsibility for members of Friedman's group. They all recalled the constant vigilance required to maintain control over their photon counters. Byram's sense of "mind over Geiger tube" was viscerally demonstrated by Robert Kreplin, when, having been asked how the delicate tubes had been

Talbot Chubb described the 1950s filling system while manipulating imaginary dials and knobs. [Still image from SVP video footage, NRL video crew].

Robert Kreplin instinctively tapped the side of a tube that had a window at one end, indicating that the anode was not supported at both ends. [Still image from SVP video footage, NRL video crew].

made sufficiently rugged to survive launch, he instinctively tapped a tube that had an anode supported only at one end.

CONCLUSIONS

Impressions I gained through archival research and through oral history interviewing have been greatly strengthened by my subsequent videohistory efforts. There is little question that these pioneer rocket groups constituted a tool-building culture defined by a set of shared experiences and institutional priorities. Without video, I still might have collected much of the same reflective commentary, but it could not have provided the same vivid surrogate experience of life in the laboratory as enacted by those interviewed, and it would not have been as detailed a study of the detectors themselves.

The videohistory sessions confirm the impression that typically a group leader looked outward to establish an agenda for the group, to announce and interpret results, and to maintain support. Friedman's role as spokesman for his group was repeated during the group sessions, whereas those who had labored at the benches focused on tools and techniques. After years of this, "testing, testing, testing" became indelibly planted in their view of life along with the spatial positions of valves and gauges on laboratory equipment replaced long ago. Compulsive fidgeting with the equipment, unwavering attention to the detectors, revealed by similar behavior during the interviews, and a shared feeling that the old days were comparatively free from bureaucracy, all strengthen the observation that most of the pioneers in these rocket teams were indeed heavily insulated from outside pressures, and indeed were left to concentrate on making things work on a rocket.

Charles Johnson's body language clearly reflected old battles as he thought back over his years of building vacuum systems for testing and evaluating his spectrometers. But these battles were with equipment, not people.[25] The interference Charles Johnson best recalls was of the radio frequency type, as he stood in a special copper-clad room at NRL that was created to isolate his experimental devices from the noisy Washington communications atmosphere.[26] A video tour of this sealed room revealed not only the degree of isolation required to test his devices, but the level of support he required to get his job done.

The videohistory sessions concentrated on revealing this support structure, not from the standpoint of national security, (which has been revealed through traditional forms of documentation), but from the standpoint of getting a technical job done. Thus I always tried to ask questions about laboratory workspaces, technical infrastructure, and the freedom to focus on technical problems. With their old detectors in hand, Talbot Chubb, Charles Johnson, E. T. Byram, and Robert Kreplin recalled the valuable technical assistance from craftsmen adept at building prototypes from sketches, glass blowing shops that could make reliable glass-Invar seals, and model shops that could adapt silk stocking weaving machines to create fine tungsten wire mesh screens.[27] The artifacts themselves, interpreted by their creators for the videocamera, represented the long-past *corpus* of the tool-building infrastructure within which they were conceived.

NOTES

1. My sense of a tool-building culture derives from observations by Howard McCurdy, "The Decay of NASA's Technical Culture," *Space Policy 5* (1989), pp. 301–310.
2. Paraphrased by George Wise, "Science and Technology," *Osiris 1*, second series (1985), pp. 229–246; quote from p. 235.
3. The definition of space science used here comes from Homer Newell, *Beyond the Atmosphere.* (NASA, 1980), p. 11. However, others have correctly pointed out that the contemporary space sciences are a blend of many cultures. 'V-2' stands for Vergeltungswaffe zwei, or Vengeance Weapon-2. On the V-2, see Gregory P. Kennedy, *Vengeance Weapon 2: The V-2 Guided Missile.* (Smithsonian, 1983); and Frederick I. Ordway and Mitchell R. Sharpe, *The Rocket Team.* (New York: Crowell, 1979).
4. Much of the following material is distilled from David H. DeVorkin, *Science with a Vengeance: The Military Origins of the Space Sciences in the V-2 Era.* (Springer-Verlag, forthcoming). Full citations to archival sources are provided in that work.
5. Johnson, F. S. OHI pp. 27–28; see also Richard Tousey OHI, 1/8/82, p. 70. SAOHP/NASM.
6. XRAYS1.01:06:29:00.
7. XRAYS1.01:58:58:00–02:00:35:00.
8. A Yale physicist who had planned to develop photographic spectrographs dropped out when he determined that photographic techniques were probably not sufficiently sensitive to do the work he felt was necessary. Photon counters would do the job, but he was not equipped to build them. See Roland E. Meyerott to Marcus O'Day, 13 January 1947, Lyman Spitzer Papers, Princeton.
9. BYRAM: XRAYS1.02:00:55:00.
10. Herbert Friedman, Talbot Chubb, E. T. Byram, and Robert Kreplin VHI, December 12, 1986, 02:20:38:00, p. 34.
11. Friedman commentary, XRAYS1.TXT 02:06:04:00, p. 27.
12. Spencer OHI, p. 37rd. SAOHP/NASM.

13. Charles Y. Johnson VHI July 8, 1987, 02:50:30:00, p. 22 Naval Research Laboratory Aeronomy: S-1.
14. XRAYS1, 03:45:35:00, p. 80.
15. Johnson Aeronomy VHI July 8, 1987, 02:46:59:00, p. 21.
16. XRAYS3:03:52:32:00–3.03:52:51:00.
17. Videohistory with Robert Kreplin and E. T. Byram, July 31, 1987. XRAYS3.TXT, 01:37:15:00, p. 20.
18. Built originally of glass piping and flasks because NRL possessed excellent glass blowing shops, these stations were eventually replaced by stainless steel systems when the glass shops closed down.
19. Talbott Chubb, July 8, 1987, 00:45:25:00, p. 21.
20. Ibid. 00:53:31:00, p. 25.
21. Ibid.
22. Julian C. Holmes and Charles Y. Johnson, July 30, 1987, 07:08:05:00 [Holmes draws a diagram on the flip chart of the dynode chain for the photomultiplier] p. 36.
23. NRL Notebooks that serve as good examples include those of F. S. Johnson, Eleanor Pressly, Homer Newell, and J. Carl Seddon. HONRL.
24. NRL S2, Julian C. Holmes and Charles Y. Johnson, July 30, 1987, 07:18:20:00–07:20:50:00, p. 42.
25. Others, apparently, took care of that, as when his associate Julian Holmes created a fictitious agency office called JPAC to cut through procurement red tape at Fort Churchill. JPAC stood for "Just Plain American Citizens." Julian Holmes oral history, NASM.
26. Videohistory NRL S-1 Charles Johnson and Robert Doggett, July 8, 1987. 03:18:52:00–03:21:41:00, pp. 38–42.
27. There was not universal agreement that the NRL shops were capable of meeting every need, especially when there were severe time pressures for meeting launch dates. Richard Tousey noted in several oral histories that he opted to have the mechanical frames for his first spectrographs built by an outside contractor, although much of the exotic optical elements were fabricated in Navy laboratories. Tousey oral history, NASM.

RECORDING VIDEOHISTORY:
A PERSPECTIVE
Brien Williams

Ours is the first century to be documented by the moving image. Engrossed as we are in our culture, it is difficult to sense fully the impact of this statement. We accept the presence of moving images as part of our reality. Yet, from the perspective of future generations, ours will be the first century for which people convey the sights and sounds of their lives to the future in the form of moving film and electronic recordings. We may be tempted to believe that our times will be better understood in the future because of this record of ourselves. In large measure this is true. But it is useful to remember that a great deal of the material we will preserve is in the form of Hollywood entertainment and heavily edited and editorialized news and documentary footage which may, or may not, accurately reflect our times.

Fortunately, there is growing interest among historians to apply the tools of film and video to the production of archival recordings which attempt to minimize the influences of mediation and editorial slant. Hopefully, this effort will achieve degrees of historical accuracy and attention to detail missing in other forms of media programming.

Increasingly, video has become the medium of choice for such documentation. Reasons for its selection include almost universal availability and diminishing costs of equipment. Video also has the advantages over film of being instantly available for playback, of conveniently recording sound and pictures together on tape, and of being relatively easy to edit and duplicate. The shelf life of video recordings, however, remains something of a mystery, although improvements in tape composition and storage techniques suggest this may become less of a concern.

While many videohistory projects are limited to the use of consumer and industrial grade video equipment and nonprofessional production crews because of budget limitations, the Smithsonian Videohistory Program, due to a generous grant, was able to take advantage of professional equipment and personnel to obtain recordings of high technical quality and production values. In this sense, the Program was experimental: (a) to find ways of applying professional, broadcast television techniques to the accumulation of documents of historical value and (b) to identify areas in which standard practices and principles of the broadcast industry would have to be altered or abandoned for ones more appropriate for videohistory.

I have served as the video producer/director for many of the Smithsonian projects. It has been exciting to participate in the development of video production practices suitable for videohistory. Some of the problems that confronted us at the beginning have been solved; others remain to challenge us. What follow are observations of mine as producer/director on how video production is best used and thought of as a research tool in the practice of videohistory.

MAJOR DIFFERENCES BETWEEN STANDARD VIDEO PRODUCTION AND VIDEOHISTORY PRODUCTION

While standard video production and videohistory share many common practices, it is best to start out by defining some areas in which the two are—or can be—distinctly different.

(1) Differing objectives. With standard video production, everything is normally geared toward a specific finished product. In most cases, it is a program for which there is a completed script or, at the least, a tightly designed production plan. Production consists in the main of finding and shooting the right pieces to fit into the preconceived program.

By contrast, videohistory recordings are end products in themselves. They are designed to be used as historical documents in their original, unedited form. Moreover, there are no scripts, as such, and productions are designed to explore topics at length. Usually they allow, if they do not actually invite, content to emerge spontaneously in the course of production. This kind of latitude is seldom tolerated in standard television.

(2) Shooting procedures. In accordance with the end product orientation of standard video production, shooting is done in a highly segmented and controlled manner. Action is broken down into individual shots which may, or may not, be recorded in natural sequence. Emphasis is placed on perfecting camera angles and obtaining technically flawless recordings. Action is frequently repeated until the right effect is achieved.

In videohistory, there is less need for such control. In fact, there is justifiable reluctance to interrupt events or alter their natural order unless absolutely necessary to obtain accurate documentation. Consequently, videohistory recordings tend to contain long, continuous shots and multiple camera moves within shots as events are covered without interruption in real time.

(3) Film language. Standard video production adheres to the "rules of film language" which have been developed over time by filmmakers and videographers to tell visual stories. Individual shots are constructed based on the way they fit into edited sequences. Typically, action is shot first in what is known as a "master shot" (a wide shot). Then it is broken down and repeated in a series of closeup, reaction, point-of-view, and other detailed shots. These are then edited together to make it seem as if action moves continuously through shot sequences.

Since edited programming is not a primary objective of videohistory, there is less need to conform to the rules of film language, at least as it pertains to the flow of edited images. (For exceptions see the "Competing Objectives" section below.) Actually, long continuous shots of sustained action is the preferred technique for videohistory. It allows for more natural on-camera behavior and shows action in organic form.

(4) Transparency. The process of making programming is almost universally hidden in standard video production. Cameras are not shown, microphones are hidden, production personnel remain unseen and unheard. Everything is geared to make the production seem transparent: creating the illusion for viewers that they are experiencing events directly.

For the most part, videohistory does not depend on this illusion. There is no need to obscure the production process. Indeed, it is more honest to acknowledge that it is taking place. Moreover, recordings are frequently improved when people on- and off-camera communicate directly with one another during production. (See the "Controlling the Flow of Objects" section below.)

(5) Extracting drama from events. It is a broad generalization, but most standard video productions are designed to extract human drama and emotion from events. Topic selection, structuring of events, and camera and microphone usage are among the variables employed to heighten the emotional impact of programing.

With videohistory, drama and emotional expression are not sought unless they are integral components of events being recorded. Consequently, employing techniques to heighten drama is avoided.

(6) Target audiences. For the most part, standard video programming is targeted to mass audiences and, therefore, is made to appeal to average tastes and interests. Even specialized programming typically employs devices to gain and hold viewer attention.

Target audiences for videohistory tend to be much narrower and consist of people who are likely to seek out the material because of prior interest in the subject matter. Thus, there is no need to resort to strategies for capturing and maintaining viewer attention.

RETHINKING THE ROLE
OF THE VIDEO PRODUCER

In the television industry, the producer's role varies greatly. In some cases, the producer is intimately involved in a subject area. Other times, he or she is expected to know very little about a topic being covered. Normally, a producer is guided by a script, a "program concept," or an executive producer who establishes the program parameters in detail. Frequently, all a field producer has to do is "fill in the blanks" of a script with "sound bites" (recorded statements) and "B-roll" (cutaways) that will make up the finished, edited "program package." Moreover, if there are "content experts" involved at all, except as "talent" (a curious term that designates anyone who appears on screen or is heard in the audio), they are usually consultants to scriptwriters and are rarely found at shooting locations.

My role as producer/director in a videohistory project has been substantially different from the industry standard. In the first place, it has been essential for me to become familiar with the subject being studied. Too many important decisions have to be made—many of them instantaneously in production—for me to be only tangentially aware of subject matter.

Secondly, I have worked very closely and continuously with a content expert, the "investigator," or historian, with whom I develop and execute a production plan. An investigator is an unknown character for most television producers. In practice, the investigator is part client, part instructor for the subject being covered, part coproducer of the project, and all host and interviewer. I learned quickly that a successful project is dependent on the establishment of a smooth working relationship between the investigator and myself. If we are not in close communication with one another and if our ideas do not coincide by the time we go into production, the results of our efforts are likely to be disappointing.

Moreover, while I have a great deal to learn from the investigator about content, in most cases I have something of a teaching responsibility of my own. Most investigators with whom I have worked lacked prior experience in video production. Yet, they had to make informed decisions with me about how to shoot, to say nothing of their having to perform on camera. Thus, one of my functions has been to provide investigators with a short course in video production, ranging from basic information on production practices to subtler details of video performance.

Another difference has been that while I was teaching investigators the common practices of television production, I was also looking for alternative ways of doing things that were better suited to archival needs. In the majority of cases, the production practices we have adapted have been hybrids, borrowing most techniques from industry standards but deviating from them in a few, sometimes substantial, ways, the most prominent of which I have mentioned above.

Whenever we have deviated from standard practices, I have also had to explain our approach to professional crews. In fact, I developed the practice of making one of my first functions with a crew the explanation of the nature of videohistory and the ways in which our production was likely to be unusual for them.

THE PROBLEM OF COMPETING OBJECTIVES

In its purest sense, the objective of videohistory is to accumulate clear and accurate audio and visual recordings of subject matter. When properly used, video can show process and place, reveal personality, and capture nonverbal language and group interactions. If this were all, the task of the video producer would be straightforward.

In practice, however, a secondary objective is almost always involved. It is the potential or actual plan to use some of the footage in an edited format, such as in an exhibit display or broadcast documentary. In some cases, the investigator has a specific objective in mind. Other times, future uses are expected but not clearly defined. Nearly always, however, we have anticipated that the subject matter of a videohistory is of sufficient interest that it will eventually find its way into some form of edited programming.

From the video producer's standpoint, there is a world of difference between a video recording that captures the reality of an event for the historical record and one that captures it in such a way that it will fit seamlessly into an edited package. Historical accuracy alone dictates one set of procedures; historical accuracy which also encompasses the needs of visual narrative (film language) is an infinitely more complicated assignment.

Whenever these objectives occur together, a major challenge of production design and execution is finding the right balance between them. A strictly archival approach requires little in the way of visual finesse. So long as the camera captures pertinent images in clear form, the historical record is served. But the same images may present serious problems to a future documentary filmmaker. Camerawork may be shaky. Picture composition may be unbalanced. Extraneous background activities may distract. From the historical standpoint, there may be no justification for stopping the flow of an event to correct such conditions. But, if the needs of a documentarian are also considered, there is every reason to make corrections. Moreover, there may be justification for adding shots to a production which have little or no historical value but are of great use to a documentarian, such as reaction shots of an investigator or wide shots of participants in a group situation.

Overall, it has been my experience that the historical purpose dominates, as it should. After all, we would not be calling it "videohistory" if our primary objective were not the historical record. I frequently deal with the suspicion, however, that in doing so we are serving a future documentarian poorly and that we ought to strive continuously for ingenious methods to meet competing ends without undue compromise.

In this connection, I have incorporated several practices into production which are minimally disruptive to the historical record and may be useful to the documentarian. Here are the most common three. First, I try to obtain at least one wide shot of an interview setting for use as an establishing shot or as a cutaway. Commonly, I ask talent to remain seated and to continue talking for a few minutes at the conclusion of an interview while I record them in a wide shot. Second, I ask the cameraperson for varying shots of the interviewee during an interview—medium shot, closeup, tight closeup—for picture variety and as a means of minimizing jump cuts among edited sound bites. I rarely ask for a noticeable change in picture composition, however, during interviewee responses because shots that are zooming are hard to handle in edited form. Third, I seek to minimize handheld shots and all shots that contain instabilities. If I see shaky camera motion, I ask for it to be done over again, unless doing so seriously interrupts the flow of an event occurring before the camera. Often, I will take time to reshoot shaky segments later. Overall, I try to limit handheld camera work to subjects and events that do not lend themselves to fixed camera coverage.

THE ADVANTAGES OF LEADING WITH AUDIO-ONLY INTERVIEWS

One of the best ways to prepare for videohistory is to conduct audio-only interviews in advance. Such interviews allow an investigator to explore topics at length with interviewees. Compared with video, audio tape and equipment are inexpensive and easily handled. Only the investigator and interviewee need be present. They can cover material at a comfortable rate in an unchallenging environment.

Going over topics in advance gives both parties the chance to "rehearse." Often reviewing something once improves a second telling. Details may be recalled in the interim. Tangential issues can be discarded. Artifacts may be identified and collected for incorporation in the video. The word "rehearse" may be troublesome to some. It is used here to denote "going over things" rather than "to practice one's lines and actions with precision." It is unlikely that a legitimate videohistory project would call for the latter practice. Also, I would not advocate advance interviews if the objective of a videohistory were specifically to capture unrehearsed responses.

Another advantage of conducting audio-only interviews in advance is that they give the investigator and interviewee the opportunity to become accustomed to interacting together in a recording situation. This experience is likely to put them at greater ease when confronted with the more intimidating video environment which includes camera, microphones, lights, and production personnel.

The investigator typically has the opportunity to review audio tapes before video production begins. This may be helpful to establish interview techniques and to select content for videotaping. In fact, leading with audio can be a crucial step in designing the video production plan and can result in significant savings of effort and expense. If the "whole story" is explored expansively in audio recordings, video may be reserved primarily to *illustrate* visual points and *complement* the audio record with visual support. As examples, videotape may be used to show a process that is difficult to describe fully by words alone, and it complements the spoken record by revealing visual aspects of personality—nonverbal expression—impossible to capture on audio tape.

Sometimes it is impractical to conduct audio-only interviews before videotaping. In such cases, time should be spent by the investigator at least going over topics in advance with interviewees, preferably on site but, if that is impossible, by phone or in written form. Such contact starts interviewees thinking about topics, accustoms them to the tone and direction of a videohistory project, and provides them and the investigator with useful information to plan the videotaping.

As a note of warning, it has been my experience that if prior contact with interviewees is slight, a great deal of the video interview will be taken up with coverage of basic information, typically presented in a disjointed fashion and much of it without visual interest. It is hard to justify the expense and effort of video production to obtain such results.

PLAN FOR THE VISUAL AND DIRECT

As a general rule, I think there is a tendency to underestimate the impact of the video image and the direct and immediate experience it can convey. It is probably natural for historians to accentuate verbal information. Videohistory challenges them to think visually and to convey visual experiences directly. Even when this is done, however, the common choice is to accompany visual information with verbal commentary. In many cases this is appropriate. Indeed, it takes advantage of one of video's special attributes: the marriage of sight and sound. But in some cases, commentary blunts the impact—to say nothing of historical value—of images which are accompanied only by natural, ambient sound. I fear some investigators regard such footage as "unexpressed information" and, therefore, of little value. To the contrary, such footage puts viewers in a direct relationship with the raw data. It allows them to experience material directly; to imagine themselves in the situation depicted. I think, too, the value of such direct experience increases over time, particularly if direct access to the raw material becomes impossible. Commentary, on the other hand, even of the best kind, inevitably functions to distance viewers from raw data by giving that data a point of view and cognitive significance.

There is another aspect of video's immediacy to consider. Often, videohistory projects dwell on past events. Interviewees share recollections, show and describe photographs, and demonstrate and operate equipment. In other media, such as print and audio, the beholder is naturally drawn back into the past along with the historical narrative. But because video is so immediate and the image so dominant, it is impossible to abandon awareness of the present even when being presented with information from the past. Whether one recognizes it or not, the most compelling component of videohistory recordings may be the present moment of the recordings themselves, rather than any recollections they contain. At the very least, the visual image serves to remind the viewer that the past is being recreated in the recorded present.

Why is this observation significant? It suggests a richness of video as a recording medium: the observed present and recollected past are presented on screen simultaneously. We watch with fascination, for example, the elderly black woman recount the story of her being denied a pilot's license as a youth because of her race (Smithsonian Videohistory Program on Black Aviators). We join her as she recreates in our minds her bitter disappointment. At the same time, we observe her philosophical resignation as it is expressed in her face. What is more, only a moment earlier we saw a picture of her in her twenties, dashing and glamorous. That image enriches her narrative and reverberates compellingly with her appearance on screen as an old person. With her and for ourselves, we shift from present to past and back again, all instantaneously as we are presented with the recorded video event.

The impact of video's immediacy should invite the selection of topics for videohistory which take full advantage of that characteristic of the medium. Among them are topics rich in visual content and those which accentuate a slice of life approach to subject matter for which considerable direct experience is available (a day in the life of robotics engineers, for example, from another Smithsonian videohistory).

ATTACHED VERSUS FREE COMMENTARY

I define "attached commentary" as that which occurs when someone is talking about a shown object or image. It may be something the interviewee and investigator stand in front of or something one of them handles. Or it may be something the camera looks at while participants discuss it off-camera. Free, or "unattached commentary," does not relate to shown objects or images.

As a general principle, when it is well handled from a technical standpoint, attached commentary results in successful videohistory, taking advantage, as it does, of video's visual impact. It brings expert and object together. Points are firmly rooted in visual data. Free commentary, on the other hand, is not so rooted. Nevertheless, the use of video to record free commentary is still justified to the extent that it reveals personality and shows nonverbal communication. Lengthy free commentary, however, may eventually lose its impact and become boring—a "talking head" interview which most of us profess a strong wish to avoid.

One frequent pitfall involves setting up a shot for attached commentary and then having the interview shift into unattached discussion. In such cases, the picture composition has been designed to accentuate the connection between interviewee and object, or the object itself, but when the commentary changes, the shot loses its jus-

C. Alfred Anderson, Janet Harmon Bragg, and Lewis Jackson were among the aviation pioneers interviewed for the Black Aviators Videohistory Series. Ms. Bragg was denied her pilot's license because of her race and her sex. [SVP photo, Terri A. Schorzman].

tification and the cameraperson is forced to make difficult, possibly unbecoming, adjustments on the run. Does the camera stay on the object previously discussed? Does the cameraperson guess what is likely to be the next portion of the object discussed and move on to it? Or does the camera come off the object and shoot the talent, who may or may not be well lit or in good position for pickup? Best is to stay with a game plan that separates attached and unattached components.

CONTROLLING THE FLOW OF OBJECTS

Integrating objects into the interview situation presents several problems. First, a decision has to be made when to have the camera move to the object and when to stay on the interviewee. Usually there is limited opportunity for rehearsing these transitions beforehand.

Objects are referred to within the context of ongoing discussion and it is difficult to predict the exact timing of their occurrence. Consequently, camera movements usually become judgment calls made by the cameraperson, unless the producer or investigator prompts them.

A related problem involves coordinating the images shown with attached commentary. When the camera shifts from an interviewee to an object, or back again, or among details within an object, the cameraperson must almost always zoom into a tight closeup to set focus, and then zoom back out again to obtain an appropriately composed shot. This process takes time—up to ten seconds or more depending on circumstances. Frequently, the commentary attached to an object is completed before the cameraperson has obtained the shot. Sometimes it has already moved on to another topic. Then the cameraperson must play catch-up, which is impractical. A similar problem occurs when an interviewee or investigator points to a detail within an object and then fails to remain pointing until the cameraperson has obtained a properly focused and composed shot of it.

There are several strategies for dealing with these situations. The first is simply to let talent know what the problems are. People will try to adjust, although it has been my experience that most find it difficult to prolong discussion or continue pointing at details long enough.

Another strategy is to have the talent and cameraperson openly coordinate the timing of object shots and commentary. Invite the cameraperson to speak up and ask the talent to delay moving on to another object or resume pointing to a detail. Encourage the talent to ask the cameraperson if he or she needs more time to obtain a shot. Such communication obviously breaks the illusion of transparency but it ensures the establishment of intelligible relationships between pictures and commentary.

Another effective technique is to have the investigator control the flow, particularly when that person has become experienced with videohistory. He or she can refer to an object which cues the cameraperson to shoot it and can continue to point to it long enough for the cameraperson to get a shot. Even better, a skilled investigator can make smooth transitions to and from objects which eliminate discrepancies between what is being said and what is being shown. For example, an investigator might say, "We want to look at this picture and identify who is in it, but tell me first where it was taken?" As the interviewee describes the place, the camera moves into a closeup of the photograph and is ready for the identification of individuals which will follow. As another example, an investigator might say, "I am pointing now to this part of the engine. While the camera shifts in on it, tell me what its function is." Such a statement cues the cameraperson to the detail and prompts a discussion of function while the camera sets up on a closeup to accompany more visually oriented commentary.

If coverage of objects has been unsuccessful during an interview, I sometimes have the camera set up to shoot them again afterwards. I may ask the interviewee, the investigator, or both, to re-identify what is being shown. Moreover, I encourage everyone to talk directly to one another—talent, cameraperson, producer—while we are shooting in order to assure that the appropriate shots are obtained and properly identified. I think of such a practice as creating a kind of visual appendix for the interview which provides clear images and unambiguous identifications.

VIDEO TOURS

In standard television production, moving camera shots are rare except in action dramas and commercials or in news stories where handheld moving shots are accepted. The vast majority of television programming is made up of edited sequences of predominantly static shots. The sense of movement comes from action within the frame and from cutting among shots.

There are at least three reasons why moving shots are not commonplace. First, they can rapidly become tiresome and distracting. Second, some camera moves are not natural. Our eyes, for instance, do not pan and zoom as the camera lens does. Third, smooth camera moves are hard to come by. To be done well, they usually require specialized camera mounts.

We have found one strong justification for using moving shots in videohistory projects. This has been when we have conducted what we call "video tours" in which a camera moves through an environment continuously recording and an expert provides commentary that identifies what is being shown.

The advantages of a video tour are that they provide a general overview of an environment and at the same time depict layout and dimensions and reveal the relationship of parts to a whole. Since the camera is often handheld in these cases, camera work may be shaky. In order to compensate for this, we usually go back through an environment afterwards to reshoot significant details with the camera mounted on a tripod.

Another successful form of "touring" has the camera moving over an object while commentary is provided as "voiceover" (speakers are heard but not seen). The camera is almost always on a tripod for this kind of tour, particularly when closeup images have to be shot from a distance. Telephoto shots accentuate the shakiness of handheld camera work.

We used this technique effectively twice while covering large telescopes. The camera started on an agreed upon shot of a portion of the instrument and then moved from element to element as prompted by the commentary. By this means, the entire telescope was shown and explained. Of course, this type of tour required close coordination and open communication between interviewee and cameraperson. Also, the commentator had to be provided with a monitor in order to observe the camera's output.

ENACTMENTS

There are times when circumstances make it difficult to record events and activities in their own reality. In some instances, we resorted to what we call "enactments," the re-creation of an event or activity for the camera.

Two examples come to mind. One involved an astronomer who uses an optical telescope. We wanted to show his normal observing techniques but they occur in darkness at night. It would have been impossible to get useful video images without resorting to specialized and expensive night vision equipment, possibly to little effect. Consequently, we asked the astronomer to go through the motions of his normal nighttime activities while we recorded him in daylight.

The second example involved documenting the operation of a pulley system used to remove slate and debris from an open pit quarry. Although the system had not been used since the 1970s, it was still operative and the investigator had been able to locate workers who used to run it. In addition to having the workers explain and demonstrate how the system worked, we wanted to get shots showing the system as if it were in normal use. We thought that since the system was obsolete and about to be torn down—and it would never fit into an exhibit space, even if it were preserved—ours was a last opportunity to show how it looked and sounded when in daily operation. Consequently, we recorded the workers in a series of individual shots which, when edited together, would depict in a smooth and logical fashion how slate had once been removed from the quarry by pulley.

How can one justify manipulations of this sort when the objective is an historically accurate documentation of process? A key, it seems

Charles Worley "enacted" a nighttime observing session during the day. [SVP photo, Terri A. Schorzman].

The cableway (pulley) system removed slate from the quarry. [SVP photo, Phillip R. Seitz].

to me, is to be honest. The videotape itself should make clear that enactments are taking place. Sometimes the process is self-evident. People are seen being given cues and are clearly observed "performing" certain acts. But if there is likely to be any confusion on the point, the investigator should explain on videotape what is happen-

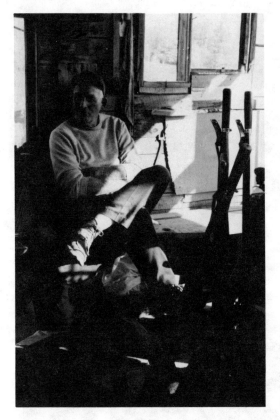

Everett Beayon returned from retirement to Vermont Slate to demonstrate the operation of the cableway system. [SVP photo, Phillip R. Seitz].

ing. In addition, hardcopy documentation that accompanies recordings should make clear that enactments have taken place and explain the circumstances.

In the examples cited, our manipulations were actually twofold. One, we asked subjects to perform actions for the camera as if they were being done "in real life." "Remember," I called out to the astronomer before a shot, "whatever you do, don't look at the camera." Such eye contact breaks the illusion of the medium's transparency ("there is no camera here, only life in the raw"). Second, we broke the subjects' actions into series of separate shots in order to show detail and optimize picture composition. When edited together, they should appear to cover continual, organic action, following the principles of film language referred to earlier.

Actually, in the examples cited, lengthy shots of continuous action seen from a single viewpoint might have been historically more accurate but less visually meaningful. Much of the astronomer's equipment is small. In order to see it and how he operates it required carefully set up detailed shots which would have been difficult to obtain on the run. On the other hand, the pulley system at the quarry is huge. No way would it have fit into a meaningful single shot. We had no choice but to break its depiction down into detailed shots: in the pit itself, at the rim, on top of the trash heap, and inside the engine house.

THE REALITY QUESTION

The primary objective of videohistory is to accumulate accurate historical records. I pointed out ways in which this objective is challenged by competing needs, particularly those of edited programming. There is another aspect of the reality question worthy of attention.

I find it useful to approach videohistory with this assumption: that the reality video captures is always one that includes the recording process in its totality. That is to say, whatever happens in front of the camera includes the camera in its consciousness. An interview between a subject and investigator is never just a conversation between two people. It is always an engagement which incorporates the presence and purposes of the camera within its reality.

It may be possible for participants in audio-only interviews to overlook the presence of the recorder once they have become accustomed to it. But I doubt it. After all, a situation in which one person asks all the questions and the other does all the responding is not a normal "real life" conversation. Even the audio-only interview involves acknowledgement of the recording process, although it may be at barely conscious levels. This is even more true in videohistory where it is impossible to ignore the recording process because of the equipment and personnel on hand.

I would go further. I believe conduct in an interview is influenced at some level by a subject's familiarity with the television interview as a genre. Behind everyone's participation is his or her own version of what they think their television "performance" ought to be. I have seen it happen too often for it to be otherwise: people undergo subtle adjustments of character and manner as videotaping begins. Such changes cannot be simply attributed to nervousness. It seems to me everyone becomes their own version of "a TV guest." Relaxed people may become energetic; tense people, laid back. The soft-spoken find uncustomary volume; boomers suddenly whisper.

After a time, of course, people tend to "settle down." In my mind that does not mean they forget about the camera's presence. Rather, it suggests to me they have adopted a "videohistory persona" with which they are comfortable. That persona is an accommodation of theirs to the recording situation and not simply a resumption of "normal" behavior. I say video reveals personality. What I really mean is that video reveals personality as it is expressed within, or in response to, the recording situation.

DIFFICULTIES PREDICTING WHAT THE FUTURE RESEARCHER NEEDS

I am fascinated by the guesswork that lies at the heart of videohistory projects. How do we cover subject matter in a way that will address the needs of researchers looking at our footage at some point in the next century or beyond? I suspect the guiding assumption is—and probably has to be—what interests us will also interest them. Yet, I am bedeviled by the fear that in failing to linger a little longer on an action or failing to pan just a little further in a shot, we deny an indistinct figure of the future the one bit of information he or she most desperately seeks.

What may appear commonplace to us and hardly worthy of attention may acquire significance over time. I am reminded of the common experience when looking at old photographs that people's outmoded dress and antiquated cars take on a fascination about which the photographer was probably unaware. Or the comment made to me recently that computer equipment of the early 1990s may leap out of the picture frame in twenty-five years because of its ridiculous bulk and the reverent attention people pay to it.

It is important to remember that our productions reflect what an investigator, interviewee, producer/director, cameraperson, and sound engineer jointly and individually think is important at a given moment of time on a few feet of videotape. Whether that coincides with a future researcher's needs is, indeed, guesswork. Moreover, it is always possible that our handling of historical material will be of greater interest to the future viewer than the historical material itself.

TECHNICAL ASSESSMENT
AND ACADEMIC APPLICATION
Kerric Harvey

BACKGROUND OF PROJECT

In 1990 I received a grant from the Smithsonian Institution to review samples from the Smithsonian Videohistory Program (SVP) collection in Washington, D.C. My dissertation research in communications at the University of Washington had, to some extent, explored the methodological ideas underlying the SVP's techniques for collecting historical information, so I was particularly pleased at the opportunity to examine their work at first hand.

While at the University of Washington, I decided to determine if taped material collected using archival techniques might be transferable for use in a larger arena, i.e., if videohistory could be shot using techniques that characterize archival tape and still be adapted for use in a broadcast setting, specifically, inclusion in national PBS educational television. I reviewed the SVP tapes from the above-stated perspective as well as for their technical viability as candidates for integration in related educational arenas, such as educational and public television programming.

RESEARCH METHODOLOGY

I developed an analytical form for content analysis of the taped material and examined the tapes in search of visual information which might not be available to researchers using other sources, such as photo files, journals, or audiotape. The form focused on nonverbal, but communicative information, which interview subjects might display during the course of their oral history sessions. I termed this type of nonverbal information "behavioral subtext."

I looked only for occurrences of behavioral subtext taking place, without making any concerted attempt to interpret the nature, intent, or meaning of any of those behaviors, since my objective was merely to determine whether or not video tape gathered information which would be unavailable through other means of oral history documentation.

I also evaluated tapes according to:

a) framing and composition;

b) use of the medium, defined by sensitivity to the main strengths of video, including the ability to convey a sense of environment, an understanding of physical context for subjects, be they animate or inanimate, and the capacity to convey emotion and nonverbal information ("behavioral subtext") along with language-based narrative; and

c) technical quality, or the degree to which the medium's skill at capturing motion and process was capitalized upon during the course of the video documentation.

I reviewed four projects in considerable depth. I looked at the Manhattan Project and Waltham Clock series for content and technical assessment, and the Twentieth Century Small Arms Development and Vermont Slate Quarry tapes for technical evaluation only.

1. Framing and Composition: Most the tapes did a commendable job regarding the framing and composition of shots and subjects, although violations of the "180 rule," the production convention which prohibits sudden and distracting shifts in screen orientation, were particularly apparent in the early Manhattan Project tapes. Approximately twenty per cent of the total footage reviewed was out of focus or so poorly lit as to be useless in a broadcast setting. Some of the hand-held footage was extremely effective, particularly portions at the Trinity test site. In contrast, taping Los Alamos personnel against a black background appeared artificial. In a related criticism, several of the sets used during this series of interviews placed the interviewer in the most conspicuous on-camera position, suggesting that he, not the subjects, was the natural center of attention and, by extension, was also the "true" focus of the interview.

A particularly striking example of this unfortunate placement resulted from the use, in several of the Manhattan Project group sessions, of a V-shaped table, which led the viewer's eye directly to the back of the interviewer's head, rather than to the subjects who were arranged around the table's outer perimeter.

2. Use of the Medium: By and large, tapes conformed to the general expectations of video production, in terms of the way in which shots were arranged into scenes and scenes then built into visually logical sequences. Both the Vermont Slate project and the Waltham Clock series were especially good at capitalizing on the strengths of the medium, letting the subjects speak for themselves and explain what they were doing as they did it, whether "it" was fitting together the tiny mechanisms that make a clock run correctly or hitching half a ton of slate to the nineteenth century overhead cableway that pried it out of the quarry.

Using video in precisely these types of situations, to illustrate mechanical processes, to capture environmental context, and to demonstrate industrial scale, seemed to produce the most effective results,
both in terms of visual impact and in terms of potential historical
value. Those segments which failed as video (although they may yet
succeed as history) were instances in which a.) the interviewer talked
more than the subject, as was often the case in the Manhattan Project
tapes, or b.) the camera taped the subject holding up, pointing at,
or standing near artifacts which were never explained, referenced,
demonstrated, or otherwise integrated into the flow of video expo-
sition. These situations were intellectually incomplete and conceptu-
ally confused.

Problems also arose from overly ambitious camera configurations.
Frequently, during single camera shoots, it appeared as if the camera
operator (or perhaps the person directing the camera operator) tried
to cram too much visual information into a single frame, which de-
tracted from the more significant content. Several situations consisted
of scenes where three or four people stand next to some large and
unexplained piece of equipment which, unfortunately, remains mys-
terious to viewers even if the experts in the frame try to describe it
to us. We can't get close enough to see exactly what it is they are
talking about as they seek to "explain" it. This was sometimes the
case with the Vermont Slate project, and often the case with the Manhattan Project tapes.

The best camera treatments were those that combined different
shots. Long shots provided information about the environment and
conveyed to viewers the "feel" of the place; closeups and cutaways
established an emotional connection between the viewer and the
video. This technique simultaneously supplied clear visual information to complement the detailed explanations being provided by interview subjects.

Tapes which captured real and spontaneous interaction among interview subjects, such as material collected during the group sessions
with women members of the Los Alamos research community, were
by far the most memorable. An important methodological footnote
to the value of videotaping group sessions, perhaps, is the opportunity to document conflicting eyewitness versions of the same event.

Video worked particularly well within the Waltham Clock series
because the coordinating historian taped her subjects talking while
they actually demonstrated the object or the process about which they
talked. Early tapes in the Manhattan Project series, in contrast, often
pictured three people sitting at a reactor site, talking not to each

other, and not about the massive backdrop which loomed behind them, but simply answering questions put to them by the interviewer. Although this scene does provide viewers with a vague sense of historical place, there was not much to be gleaned from that setup which couldn't be documented any other way, with far less bother and at much less cost. Certain points in this session were interesting but were hampered by the awkward seating arrangement of the on-camera subjects, who were lined up facing the interviewer rather than each other. If all three subjects were arranged physically to feel more comfortable talking among themselves, or if the relationship of the camera itself to the three subjects had been less rigid and emotionally distancing, the give-and-take aspect of their conversation may have been enhanced.

3. Technical Quality: In developing guidelines for assessing the technical quality of the SVP tapes, I confined my evaluation to the minimum standards required to preserve a video signal, such as those imposed by the editing process and by the transmission of that signal through broadcast or cable channels. Specifically, I looked at signal strength (related to the absence or presence of sufficient light in the production environment), signal quality (the reliability and consistency of color representation and the amount of break-up visible to the eye during the review process), and signal stability (the absence or presence of vertical rolling, image flickering, and/or speed lines within the frame).

A full test of the material's technical integrity would require more extensive treatment and more sophisticated technical measures than I was able to apply under this particular set of research circumstances. Nonetheless, the technical quality of the SVP footage was certainly sufficient to permit its use in broadcast transmission. Only about ten to fifteen per cent of the tape which I viewed would be, I believe, technically unsuitable for use in a broadcast capacity, although based on the assessment described in the previous section, about twenty to thirty-five per cent, perhaps, would prove aesthetically problematic.

INTERVIEWS WITH HISTORIANS

I spoke with the historians who conducted the Manhattan Project and Waltham Clock series. Stanley Goldberg, historian for the Manhattan Project tapes, found that his only concern with videotaping was knowing when enough was enough; to Goldberg, video offers such a wealth of documentation possibilities that the temptation to tape everything, regardless of its historical importance, was very strong. He was enthusiastic about the utility of videotape as a tool in historical research, citing, however, the need for the aspiring video historian to reconcile "conflicting pressures to tell the truth without violating your subjects' privacy." His concern with the ethical issues pertaining to video history is an important one, which the second interview subject also discussed.

Carlene Stephens noted the important issue of researcher bias in videotaped oral history of the Waltham Clock Company by addressing the idea that the video medium may be somewhat more vulnerable than print to unwitting infiltration of the investigator's personal perspective. Choices about camera angle, framing, dramatic re-enactment, and so on, she noted, " . . . take on a journalistic (as opposed to a historical) aspect when they are executed in the video medium." Despite best intentions, video may manipulate because its seemingly representational nature permits a camouflaged level of distortion.

PRESENCE AND LEVEL OF
NON VERBAL INFORMATION

SVP material contained non verbal, visual documentation. Standard body language displays, physical interactions among participants in a group session or between the interviewer and an interviewee (for example, shaking one's fist or pointing one's finger), and the demonstration of, or physical reference to, an artifact or actual site by interview subjects are examples that come to mind.

Several spontaneous occurrences illustrate the potential heuristic value of this type of behavioral information. For example, a group of researchers and technicians who had participated in the nuclear bomb project at Los Alamos discuss the emotions they felt upon learning the news of civilian deaths at Hiroshima and Nagasaki. Their individual body language during this episode displayed well-documented indicators of stress, ambivalence, and internal conflict, which were often in direct contradiction to their spoken justifications of the use of their research efforts on noncombatants in the devastated Japanese cities.

I can make no overall claim about the relative academic value of the data concerning behavioral subtexts, but my research indicates that it is present to a significant degree. Using a set of observational measures developed specifically for application in SVP research, I noted a total of 1,982 incidents of behavioral subtext display in 362 minutes of SVP taped material, which averages out to approximately 5.47 displays per minute, or one every eleven seconds.

GENERAL COMMENTARY

Both the historian and director must recognize that archival tape is shot for a different purpose and destined for a different fate than educational TV. As such, the tolerance level for aesthetic sophistication in the types of visuals collected for archival purposes can, and should, be significantly higher than for its broadcast-bound counterparts.

Some of the most moving and illuminating SVP footage—tape which provided the most startling, whimsical, or emotionally powerful insights into other people's lives—was collected during times when the interview participants thought that the camera had been turned off. This was a rare occurrence in the SVP tapes, and could not in any way be construed as a deliberate practice exercised on the part of Smithsonian historians or producers. But even its occasional and unintentional presence raises the point that, using the video medium, it is possible to record subjects without their knowledge or their consent, and that there may arise instances in which the most exciting information gathered is that which its source never intended to share. Even more so than journalists, who wrestle daily with the ethics of information gathering, video historians must be aware of and resistant to the temptations of surreptitious or serendipitous data collection techniques.

APPENDIX 1: PROJECT SUMMARIES

The Smithsonian Videohistory Program worked with eighteen Smithsonian historians in recording twenty-two projects in the history of science and technology. A summary of projects, based on finding aids, follows.[1] Project titles include the record unit (RU) number assigned by the Smithsonian Institution Archives, which provides easy access to materials.

BLACK AVIATION PIONEERS (RU 9545)

African-American men and women struggled throughout the 1930s to gain the opportunity to fly airplanes. Despite institutionalized biases and the Great Depression, the number of licensed black pilots increased about tenfold, to 102, between 1930 and 1941. This development helped move the federal government, but not the private sector, into sanctioning black men to operate airplanes during WWII. Theodore Robinson, visiting historian at the National Air and Space Museum, interviewed five black aviators of the 1930s. These pioneers related how they became interested in flying, obtained airplanes and training, and contended with racial (and gender) prejudices opposing them.

Participants: C. Alfred "Chief" Anderson, Janet Harmon Bragg, Lewis A. Jackson, Cornelius Coffey, Harold Hurd.
Locations: National Air & Space Museum, Smithsonian Institution; Carnegie-Woodson Regional Library, Chicago, Illinois.
Sessions/Dates: 2 sessions recorded between November 1989 and March 1990.
Materials: 4 tapes (7. 0 hours) and transcripts.

CELL SORTER (RU 9554)

The fluorescence-activated cell sorter has become an important tool in the study of cell biology and immunology, and in AIDS research. It was developed at Stanford University and commercially produced by Becton Dickinson Immunocytometry Systems (BDIS) of San Jose, California. Ramunus Kondratas, curator at the National Museum of American History, interviewed several key people in-

161

volved with the development, manufacture, and application of the machine.

Participants: Bruce Allen Bach, Mack J. Fulwyler, Leonard Herzenberg, Leonore Herzenberg, Nahesh S. Mhatre, Richard Owens, Diether J. Recktenwald, M. Boris Rotman, Bernard Shoor, Marvin Van Dilla.
Locations: BDIS, San Jose, California; National Museum of Health and Medicine of the Armed Forces Institute of Pathology, Walter Reed Army Medical Center, Washington, D.C.; Bio-Med Center, Division of Biology and Medicine, Brown University, Providence, Rhode Island.
Sessions/Dates: 4 sessions recorded between January and June 1991.
Materials: 7 tapes (10 hours) and transcripts.

CLASSICAL OBSERVATION TECHNIQUES (RU 9534)

Developments in photography and nonvisible radiation technology in the late twentieth century have rendered the classical astronomical techniques of visual observation obsolete. Astronomers at the United States Naval Observatory (USNO), however, have continued to visually observe and measure certain classes of double stars whose qualities fall between the measurement capacities of photographic and photometric techniques. David DeVorkin, curator at the National Air and Space Museum, conducted videotaped interviews with one of the last "classical observers" and with his associates about the rationale for the methodology and their incorporation of newer (electronic) methods. The sessions also documented the operation of the 26 inch telescope, the 6 inch transit, the photographic zenith tube telescopes, and computerized instrumentation.

Participants: Thomas Corbin, Geoffrey Douglass, Stephen Gauss, Dennis McCarthy, Charles Worley.
Locations: USNO, Washington, D.C.
Sessions/Dates: 3 sessions recorded between March 1988 and May 1991.
Materials: 5 tapes (6.5 hours) and transcripts.

CONSERVATION OF
ENDANGERED SPECIES (RU 9553)

Endangered species are generally preserved in one of two ways, either by existing in a controlled breeding population to ensure genetic diversity, or by living in a conserved habitat to ensure the preservation of a wild population. The Smithsonian sponsors both methods, which represents a significant departure from the earlier traditions of animal exhibition and study of museum specimens in isolation. Pamela Henson, historian for the Smithsonian Institution Archives, interviewed scientists representing both schools of thought about their role in policy making, and how their views differed in managing animal populations. She also documented the sites and the interaction of the scientists with the animals.

Participants: John Christy, Larry Collins, Norman Duke, Scott Derrickson, Robin Foster, Brian Keller, Gilberto Ocana, Stanley Rand, Theodore Reed, Ira Rubinoff, Alan Smith, Nicholas Smythe, Linwood Williamson, Donald Windsor.
Locations: Smithsonian Tropical Research Institute, Smithsonian Institution, Panama; National Zoological Park, Smithsonian Institution, Washington, D.C.; Conservation and Research Center, Smithsonian Institution, Front Royal, Virginia.
Sessions/Dates: 13 sessions recorded between June and September 1990. [Also, oral history sessions with T. Reed pertain to topic and are kept with the Oral History Project, Smithsonian Institution Archives].
Materials: 13 tapes (12 hours) and transcripts.

DEVELOPMENT OF THE ENIAC (RU 9537)

The ENIAC (Electrical Numerical Integrator and Computer), the largest and most powerful early computer, was designed to compute the paths of artillery shells, and to solve computational problems in fields such as nuclear physics, aerodynamics, and weather prediction. David Allison, curator at National Museum of American History, interviewed the co-designer of the ENIAC about its design, development, and operation, including technical and non-technical aspects. The interviewee demonstrated the operation of accumulators, wiring conduits, and function tables with the original artifacts displayed in the gallery.

Participant: J. Presper Eckert.
Location: National Museum of American History, Smithsonian Institution, Washington, D.C.
Session/Date: 1 session recorded in February 1988.
Materials: 2 tapes (2.5 hours) and transcript.

DNA SEQUENCING (RU 9549)

In 1986 the California Institute of Technology announced its development of a semiautomated machine for sequencing DNA. It became a key instrument in mapping and sequencing genetic material. That same year, Applied Biosystems, Inc. produced the first commercial instruments for clinical use. Ramunus Kondratas, curator at the National Museum of American History, documented the history, development, and applications of the DNA sequencer. He also explored the commercialization of the instrument, including its testing and marketing, and addressed current and future use of the machine in medical research.

Participants: Kurt Becker, Leroy E. Hood, Michael Hunkapillar, Robert J. Kaiser, Anthony R. Kerlavage, W. Richard McCombie, Marilee Shaffer, Lloyd M. Smith, J. Craig Venter.
Locations: California Institute of Technology, Pasadena, California; Applied Biosystems, Foster City, California; National Institutes of Health, Washington, D.C.
Sessions/Dates: 3 sessions recorded between October 1988 and March 1990.
Materials: 5 VHS tapes (8.8 hours); transcripts.

INTERNATIONAL ULTRAVIOLET EXPLORER (IUE) (RU 9543)

The IUE geosynchronous satellite, launched in 1978, was the only American astronomical telescope working in orbit until the launch of the Hubble Space Telescope in 1990. Many discoveries emerged from the IUE, including the detection of sulfur in the nucleus of a comet and the observation of a massive hot halo of gas surrounding the Milky Way. David DeVorkin, curator at the National Air and Space Museum, interviewed scientists about the creation, design, manufacture, administration, and use of the IUE.

Participants: Carol Ambruster, Albert Boggess, Yoji Kondo, George Sonneborn, Charles Loomis, Lloyd Rawley, Mario Perez.
Locations: Goddard Space Flight Center, NASA, Greenbelt, MD: telescope operations control center, data collection rooms, visitor center exhibit hall.
Sessions/Dates: 2 sessions in March 1990.
Materials: 4 tapes (6.4 hours), transcripts.

THE MANHATTAN PROJECT (RU 9531)

The U.S. effort to build and explode the first atomic bomb was known as the Manhattan Project. Stanley Goldberg, consulting historian to the National Museum of American History, recorded eighteen video sessions, divided into five collection divisions, with participants involved in the engineering, physics, and culmination of the Manhattan Project. Goldberg examined the research and technologies necessary to realize the uranium and plutonium bombs. He supplemented interviews with visual documentation of the industrial plants that refined and separated the isotopes, and of the machinery that delivered and dropped the bombs. Interviewees explained the other steps of designing, building, testing, and detonating an atomic bomb. Discussions also elicited a social history of the Project as men and women recalled different duties in different locales.

Participants: Harold M. Agnew, Frederic W. Albaugh, Frederick L. Ashworth, Dale F. Babcock, Robert Bacher, Kenneth T. Bainbridge, George M. Banic, Jr., Hans Bethe, Colleen Black, Connie Bolling, Lyle F. Borst, Norris E. Bradbury, Jess R. Brinkerhoff, R. M. Buslach, Wilson A. Cease, Vivian Russell Chapman, Edward C. Creutz, Robert Christy, Lawrence Denton, Bernard T. Feld, Richard F. Foster, Anthony French, David H. Frisch, Rose Frisch, John M. Googin, Oswald H. Greager, David Hawkins, Donald Hornig, Lillian Hornig, Paul Huber, Chris P. Keim, Clarence E. Larson, Jane W. Larson, Audrey B. Livingston, Robert S. Livingston, J. Carson Mark, Franklin T. Matthias, William P. McCue, Phillip Morrison, James A. Parsons, Leonard F. Perkins, Sr., Norman F. Ramsey, Frederick Reines, Robert Serber, Alice Kimball Smith, Cyril Smith, Charles W. Sweeney, Ralph K. Wahlen, Albert Wattenberg, Alvin Weinberg, Eugene P. Wigner, Jane S. Wilson, Robert Wilson, Wakefield A. Wright.
Locations: Hanford Engineering Works, Hanford, Washington; re-

cording studio, Boston, Massachusetts; gaseous diffusion plant site, Oak Ridge and recording studio, Louisville, Tennessee; White Sands Missile Range/Trinity test site, Alamogordo and Los Alamos Scientific Lab, Los Alamos, New Mexico; National Museum of American History, Smithsonian Institution, Washington, D.C.; and Garber Facility, Smithsonian Institution, Suitland, Maryland.

Sessions/Dates: 18 sessions (5 collection divisions) recorded between January 1987 and June 1990.

Materials: 29 VHS tapes (39 hours), transcripts.

MARINER 2, TWENTY-FIFTH ANNIVERSARY (RU 9535)

Launched in 1962, Mariner 2 was the first successful encounter of a spacecraft with another planet (Venus). It returned important data on the electromagnetic and energetic particle environment of interplanetary space. Allan Needell, curator at the National Air and Space Museum, moderated a session of engineers, scientists, and administrators associated with Mariner to discuss the construction, launch, and operation of the planetary probe.

Participants: Jack Albert, Albert R. Hibbs, Lewis D. Kaplan, Jack N. James, Oran W. Nicks.

Location: Ripley Center, Smithsonian Institution, Washington, D.C.

Session/Date: 1 session recorded December 11, 1987.

Materials: 2 VHS tapes (3 hours), transcript.

MEDICAL IMAGING (RU 9544)

The ACTA (Automatic Computerized Transverse Axial) scanner was developed in 1973. The introduction of this first full-body scanner lead to advancement in medical imaging and diagnostic medicine, especially for noninvasive viewing of soft tissue inside the body. Ramunus Kondratas, curator at the National Museum of American History, interviewed the scanner's designer as well as radiology experts who use the machine.

Participants: David Griego, Robert Ledley, Seong Ki Mun, Homer Twigg, Robert Zeman.

Locations: "Medical Imaging" exhibit, National Museum of American History, Smithsonian Institution, Washington, D.C.; National

Biomedical Foundation, Washington, D.C.; Georgetown University Medical Center, Washington, D.C.

Sessions/Dates: 3 video sessions recorded between April 1989 and October 1989; 1 audio session recorded in June 1989; 2 documentaries filmed in 1978 and 1984.

Materials: 4 VHS tapes (12 hours); 3 audio tapes (4 hours); 2 film to video (VHS) transfers.

MINI- & MICROCOMPUTERS (RU 9533)

An informal confederation of computer software designers, known as "the brotherhood," formed during the late 1970s. The group began as a result of the members' mutual interest in microcomputer software development and their geographic proximity in California. Their contribution to computer graphics and games was significant in the development of more advance systems. Jon Eklund, curator at the National Museum of American History, interviewed six members of the group about creating, publishing, marketing, distributing, and reporting of microcomputing software in the late 1970s. They also demonstrated computer games.

Participants: Dave Albert, Douglas Carlston, Margot Comstock, Jerry Jewell, Ken Williams, Roberta Williams.
Location: Broderbund Software, Inc., San Rafael, California.
Session/Date: 1 session recorded in July 1987.
Materials: 3 tapes (1.6 hours), transcripts.

MULTIPLE MIRROR
TELESCOPE (MMT) (RU 9542)

The MMT was the prototype for new and radically designed astronomical telescopes. It was the world's first large-scale multiple mirror telescope, which used the combined light of six 72-inch reflecting telescopes in a single altitude-azimuth mount. Computers controlled all pointing and tracking of the MMT's individual telescopes. David DeVorkin, curator at the National Air and Space Museum, recorded six sessions with astronomers, opticians, and engineers to document the design, construction, and operation of the telescope.

Participants: Nathaniel Carlton, Frederic H. Chaffee, Craig Foltz, Carol Heller, Keith Hege, Thomas Hoffman, Aden Meinel, Michael

Reed, Robert Shannon, Ray Weyman, Joseph T. Williams, Fred L. Whipple.
Locations: MMT Observatory, Mt. Hopkins, Arizona; Channel 18 recording studio, Tucson, Arizona.
Sessions/Dates: 6 sessions recorded in May 1989.
Materials: 7 VHS tapes (8.7 hours), transcripts.

NAVAL RESEARCH LABORATORY SPACE SCIENCE (RU 9539)

Following World War II, the U.S. Navy funded studies in astronomy and aeronomy at the Naval Research Laboratory with captured German V-2 missiles. These studies resulted in more sophisticated views of the composition of the atmosphere and of solar radiation, and in the revelation of the presence of solar X-ray radiation. David DeVorkin, curator at the National Air and Space Museum, recorded five sessions, organized in two collection divisions, with men who pioneered the science of aeronomy and X-ray astronomy. The participants discussed how they adopted, applied, or improved on extant technologies for hybrid research.

Participants: Edward T. Byram, Talbot A. Chubb, Herbert Friedman, Julian C. Holmes, Charles Y. Johnson, Robert Kreplin.
Locations: National Air and Space Museum, Smithsonian Institution, Washington, D.C.; Naval Research Laboratory, Washington, D.C.
Sessions/Dates: 5 sessions recorded between December 1986 and July 1987.
Materials: 8 VHS tapes (13 hours), transcripts.

NEW UNITED MOTORS MANUFACTURING, INC. [NUMMI] (RU 9550)

In February 1983 General Motors (GM) Corporation entered into a joint venture with Toyota to produce automobiles using Japanese management techniques at a GM plan in Fremont, California. The plant was, at that time, the least productive in the GM system. The combined corporate effort, known as NUMMI, opened for production in December 1984. Within five years the plant operated as efficiently as Japanese manufacturing facilities. Peter Liebhold, museum specialist at the National Museum of American History, toured the factory and its production lines to document the mechanical applications of Japanese managerial philosophy. He interviewed employees throughout the plant for their responses to the organizational changes.

Participants: Michael Damer, Gary Convis, George Nano, many others.
Location: NUMMI, Fremont, California.
Session/Dates: 1 session recorded in September 1990.
Materials: 3 VHS tapes (5.6 hours); transcripts.

PORTRAITS: MARGARET J. GELLER (RU 9546) RESTRICTED

Margaret J. Geller, professor of astrophysics, Harvard University, and astrophysicist, Smithsonian-Harvard Center for Astrophysics, is highly regarded for her revolutionary work on the large-scale structure of the universe. The discovery by Geller, John Huchra, and Valerie de Lapparent of the bubble structure of the universe is among the most important work in recent astronomy. Matthew Schneps, codirector of the Wolbach Image Processing Laboratory at the Smithsonian Astrophysical Observatory, and David DeVorkin, curator at the National Air and Space Museum, interviewed Geller in two separate sessions. Schneps focused on early influences in Geller's life and work as a student and scientist. DeVorkin looked at Geller's scientific interests and activities to gain a greater sense of Geller's contributions to the field of astronomy.

Participants: Margaret J. Geller, Emilio Falco, Massismo Ramella.
Locations: Geller's home and office, Cambridge, Massachusetts; Smithsonian Astrophysical Observatory image-processing laboratory, Smithsonian Institution, Cambridge, Massachusetts.
Sessions/Dates: 2 sessions recorded between February 1989 and July 1990.
Materials: 4 VHS tapes (6.8 hours); transcripts.
Special Conditions: Restricted.

THE RAND CORPORATION (RU 9536)

The RAND Corporation was established in 1945 as a United States Air Force project under contract to the Douglas Aircraft Company. Its broadly defined function was to study American national security and the role of airpower in that context. National Air and Space Museum curators Joseph Tatarewicz, Martin Collins, and Paul Ceruzzi interviewed twenty-two current and former RAND employees about their work in photoreconnaissance, about research and intellectual culture, and about pioneering work in computer development. Sessions are organized in three collection divisions.

Participants: Paul Armer, Bruno Augenstein, Edward J. Barlow, Morton I. Bernstein, Barry W. Boehm, Raymond W. Clewett, Merton Davies, Edward C. DeLand, Thomas O. Ellis, Irwin Greenwald, Gabriel F. Groner, William F. Gunning, Amrom Katz, Burton H. Klein, Walter Levison, William P. Myers, Robert T. Nash, Keith W. Uncapher, J. Clifford Shaw, Gustave Shubert, Robert D. Specht, Hans Speier, Willis Ware, Albert Wohlstetter.

Locations: "Looking at Earth" exhibit gallery, National Air and Space Museum, Smithsonian Institution, Washington, D.C.; private home and RAND headquarters, Santa Monica, California.

Sessions/Dates: 8 sessions (3 collection divisions) recorded between January 1987 and June 1990.

Materials: 15 VHS tapes (19.7 hours), transcripts.

ROBOTICS (RU 9552)

Robotics is the art of intelligent machines, a field of research that combines electrical, electrical, and mechanical engineering. Steven Lubar, curator at the National Museum of American History, recorded sessions with robot designers at two universities and one private corporation to document different work styles, environments, and the processes by which engineers make decisions. He also documented teamwork, robots in use, and the interaction between people and robots.

Participants: Brian Albecht, Shapour Azarm, John Bares, Steve Bartholet, Jigien "Roger" Chen, Kevin Dowling, Robert Drap, Regis Hoffman, Eric Krotkov, Henning Pangels, Joel Slutzkey, Armen Sivaslian, Peter S. Tanguy, Lung-We Tsai, David Wettergreen, William "Red" Whittaker, and many students from the University of Maryland.

Locations: Department of Mechanical Engineering, University of Maryland, College Park, Maryland; Odetics, Inc., Anaheim, California; Field Robotics Center of The Robotics Institute, Carnegie-Mellon University, Pittsburgh, Pennsylvania.

Sessions/Dates: 4 sessions recorded between March 1989 and September 1990.

Materials: 7 VHS tapes (11.0 hours); transcripts.

SMITHSONIAN INSTITUTION PALEONTOLOGY (RU 9530)

The National Museum of Natural History (NMNH) of the Smithsonian Institution houses one of the world's major paleontological collections. In addition, curators there have developed many innovative techniques for handling, processing, and interpreting fossils. In these video interviews, Pamela Henson, historian for the Smithsonian Institution Archives, used the fossil collections to stimulate discussion of the history of the collections and to visually document fossil preparation techniques.

Participants: G. Arthur Cooper, J. Thomas Dutro, Jr., Richard E. Grant, Ellis L. Yochelson.

Locations: Collections areas, library, office, and acid etching room in the National Museum of Natural History, Smithsonian Institution, Washington, D.C.

Sessions/Dates: 3 sessions recorded between May 1987 and August 1988.

Materials: 5 VHS tapes (4.2 hours); transcripts.

SOVIET SPACE MEDICINE (RU 9551)

The Institute for Biomedical Problems was founded in 1963 to undertake the study of space medicine. It is located in Moscow, Russia, and consists of a Primate Space Flight Training Center, research laboratories, and a small museum. Cathleen Lewis, curator at the National Air and Space Museum, interviewed scientists about their research and participation in the Soviet aviation and space medicine program prior to 1964, and at the Institute. Lewis documented early work in the fields of aviation and space medicine as well as museum exhibits about the Institute's work in space exploration.

Participants: Oleg Gazenko, Abraham Genin, Irina Gireeva, Vladimir Magedov, Eugenii Shepelev.

Locations: Institute for Biomedical Problems, Moscow, USSR.

Sessions/Dates: 5 video sessions and 1 audio session recorded in November 1989.

Materials: 5 VHS tapes (8.4 hours); 1 audio tape, (1.25 hours); transcripts.

TWENTIETH CENTURY
SMALL ARMS (RU 9532)

Technological and organizational developments changed the way military small arms were designed in the second half of the twentieth century. The use of alloys and composite materials required more specialized knowledge than one person could master, and bureaucratized weapon procurement policies required a corporate group to finance and represent innovative weapon concepts. Consequently, it became almost impossible for one person to design, build, and market a new small arm.

Edward C. Ezell, curator at the National Museum of American History, recorded twelve sessions, organized in three collection divisions, with two of the last solo designers in the world. Ezell was interested in the process by which the men developed and producer their designs, and in their experiences with respective military bureaucracies.

Participants: Eugene M. Stoner (M16 automatic rifle), Mikhail T. Kalashnikov (AK47 automatic rifle).
Locations: Ares, Inc., Port Clinton, Ohio; various sites in Moscow and Leningrad, Russia; Northern Virginia Rod and Gun Club, Star Tannery, Virginia.
Sessions/Dates: 12 sessions (3 collection divisions) recorded between April 1988 and May 1990.
Materials: 14 VHS tapes (16.8 hours) and transcripts of each session, including Russian/English translation.

VERMONT STRUCTURAL SLATE (RU 9547)

The Vermont Structural Slate Company in Fair Haven, Vermont, was founded in 1859. As of 1989, it was one of twenty remaining companies nationwide that produced slate. William Worthington, museum specialist at the National Museum of American History, recorded quarrying methods to document remaining nineteenth century industrial techniques before installation of more modern equipment. He also documented various methods and equipment used in making slate shingles.

Participants: Brad Bauman, Everett Beayon, Joseph Root, Raymond Cull.

Location: Eureka quarry, roofing shed, rubbish heap, and general operations of Vermont Structural Slate Company, Fair Haven, Vermont.

Sessions/Dates: 2 sessions recorded in October 1989.

Materials: 3 VHS tapes (4 hours); transcripts.

WALTHAM CLOCK COMPANY (RU 9548)

Waltham Clock Company was founded in 1850. Employees pioneered the machines and techniques necessary for the mass-production of pocket watches that made the company the dominant American watch manufacturer in the nineteenth and early twentieth centuries. By the late twentieth century Waltham Clock was one of the last firms in the United States still producing mechanical watches. Carlene Stephens, curator at the National Museum of American History, documented operating machinery and recorded the process of making and testing mechanical watches.

Participants: Marie Bastarache, David Buccheri, Richard Halstead, Stanford James, Tam Thi Le, Bruce LeDoyt, Joseph "Chuck" Martin, Edward Murphy, Charles Paradis, Edward Pitts, Vincent Rhoad, Richard Welch, Savay Xayavong.

Location: Secondary operations, machine shop, testing operations, final assembly at the Waltham Clock Company, Waltham, Massachusetts.

Sessions/Dates: 2 sessions recorded in June 1989.

Materials: 4 VHS tapes (5.2 hours); transcripts.

NOTES

1. Summaries are based upon finding aids written by Joan M. Mathys, Alexander B. Magoun, Laura Kreiss, Maureen Fern, and Terri A. Schorzman. *The Guide to the Collections of the Smithsonian Videohistory Program* provides a full description of all projects; it can be obtained by contacting the Smithsonian Institution Archives, MRC 414, Washington, D.C., 20560, 202-357-1420.

APPENDIX 2: WORKSHEETS
AND SAMPLE DOCUMENTS

I. GENERAL INFORMATION

1. OUTLINE OF USEFUL INFORMATION

A videotaped oral history project, if planned with care, can produce rewarding results. The medium, however, can be time-consuming and expensive. The following outline lists examples of technical, logistical, and archival information needed for undertaking a videohistory project.

Recording Formats

There are a number of formats used for videotaping in the United States. Formats include 1″, Betacam, 3/4″, VHS, and 8mm (video 8), among others. The size of stock and speed of recording will determine technical quality and longevity of the product.[1]

I.   Professional Quality

    1) 1 inch: [1″ Type C] High quality broadcast and studio video tape format. Open reel.

    2) 3/4 inch: [Sony U-Matic]. Used widely for broadcast, industrial, and ENG. Cassette only. Played on U format machines only; not compatible with other cassette players.

    3) Betacam: High-speed 1/2″. Broadcast quality. High image quality. Cassette.

    4) Betacam SP: Newer version of Betacam. Improved picture resolution and dubbing capability. Cassette.

II. Consumer

    1) VHS: [1/2″ cassette] Not interchangeable with Beta format. Cassette. Dominant throughout world.

    2) S-VHS: [Super VHS] Better picture quality. A possible replacement for 3/4″ U-Matic in the industrial and high-end consumer market. Cassette.

    3) Betamax: [1/2″ cassette]. Not interchangeable with VHS format. Cassette only. Limited use in U.S.

4) 8mm Video: [Video 8] Amateur videotape format intro-
                duced by Sony in 1984. High 8, introduced in
                1989, offers near-broadcast quality.

Equipment and Personnel

Equipment used will depend on recording formats, field or studio
production, additional technical requirements, etc. Complexity varies
with taping session. The following are elements in most video set
ups. A video shoot also requires more people than associated with
an audio taped oral history, and can include a technical crew, director, and assistant.

I. Equipment

1) Camera and Recorder/deck: Records audio and visual sig-
                               nals.
2) Microphone: Picks up sound. Separate microphones, such
                 as lavalieres and booms, should be used.

3) VCR/VCP: Video Cassette Recorder; will record and play
              back tape. Video Cassette Player; plays but
              does not record.

4) Switcher: Cuts electronically from one source to an-
               other; from one camera to another in a multi-
               camera setup, for example.

5) Lights: Direct or soft/reflected; corresponding equip-
             ment.

6) VTR/VTP: Video Tape Recorder. Will record and play
              back tape. Video Tape Player. Plays but does
              not record.

7) Monitor: Television screen or receiver that displays vid-
              eotaped material. Director and/or interviewer
              can use it to view image being recorded.

8) Camcorder: Camera and VCR in one shoulder-weight unit.

9) EFP: Electronic Field Production, uses broadcast
          standard equivalent of the camcorder; usually
          a single camera with the capacity to record on
          location or input live into broadcast (field re-
          cording). Often the best way to record video-
          history.

II. Personnel

1) Administrator: [Executive Producer] Oversees completion of all projects, and sessions within projects; handles budgets, contracts, and tracks all paperwork; works closely with historian and director to evaluate session requirements (content, visual, and technical); recommends and hires directors and occasionally crew.

2) Producer: For videohistory, the producer is often the historian/interviewer, since s/he selects the topic and interviewees; organizes the sessions; schedules dates; determines shoot locations appropriate to topic; selects artifacts; works with director to determine how best to capture "visual information."

3) Director: Coordinates technical aspects of shoot and ensures that audio and video meet standards; (often) subcontracts and supervises crew; selects shots and instructs camera operator; helps historian become comfortable with medium. Historian/interviewers should rarely, if ever, take on this position, unless they understand the mechanics of a shoot and work with a crew they know and trust.

4) Crew: The *camera operator* should be very experienced in composing shots, in following action, in listening to the content (to anticipate where to shoot); often follows guidance of director, but often relies on own discretion, particularly when in the field. The *sound operator* determines microphone requirements; sets up equipment; mikes people; determines and monitors audio recording levels and timecode; changes tape; notifies interviewer when tape runs out. May also handle most of the equipment for set-up and tear-down, unless a *grip* is available (an assistant). Most crews will not let the historian/interviewer, and often the director, touch the equipment.

5) Production Assistant: For a videohistory shoot, the production assistant provides essential service. S/he creates tape log (subject matter, place, date, etc); gets release forms signed; sketches production set-up; notes problems and solutions; provides all documentation for shoot; hand carries master tapes to office for formal labeling, and prepares for duplication.

Planning

Planning a video taped oral history project is extremely important—and is more complex than arranging an audio recorded interview. [See Planning and Assessment regarding questions about when and why to use video]. A videotaped interview involves more equipment and people than does an audio interview; and the cost is often prohibitive to many small programs. Archival processing (duplication and transcription) is also more expensive.

1) Budget: How much money is available; what can be done given the amount? (i.e. low-level equipment, non-professional technicians for many sessions, or high-level equipment and professionals for a limited number of sessions?

2) Time-frame: Deadlines, time constraints.

3) Research: Know the subject matter.

4) Location: Determine lighting quality, noise level, set-design possibilities and contribution of environment to quality of video.

5) Artifact selection: photos, slides, objects, equipment, environment; select items that add visual meaning and texture to interviews.

6) Production:
   a) Producer: professional help determines set-up, design, technical and equipment requirements.
   b) Crew size: depends on complexity of shoot, # cameras, etc.
   c) Time code: time reference recorded on spare track of tape.

d) Studio v. remote: in a studio setting or in the field.

e) Equipment: dollies, lights, microphones, cameras, decks, etc.

f) Participants: usually the subject of production, the interviewees

g) Contract: specify requirements before production begins

6) Checklists: ensure that all tasks are completed before, during, and after the shoot;

7) Logistics: vital information, the "who, what, when, where, and why;"

8) Release forms: participants sign over their share of copyright to make recorded material available for others;

9) Copyright: guides broadcast and duplication rights; put copyright at beginning of tape;

10) Tape identification: who, what, where, when, tape generation, project name.

Processing

"Processing" refers to care and maintenance of completed video sessions, including technical, archival, and storage concerns. For storage purposes, tape must be housed properly to assist the longevity of tape. The following guidelines suggest ways to preserve tape, although ideal conditions do not, as yet, exist.

1) Duplication: video can be transferred from any one format to another. The ease of duplication raises security and copyright issues.

   a) master to dubbing master (master to 3/4″ U-Matic);

   b) dubbing master to use copies (VHS) with time-code windows;

   c) audio cassettes (for transcription).

2) Log or index.

3) Name and word lists.

4) Transcription (verbatim).

5) Audit check, copy edit, visual-cue, time-code cross reference.

6) Finding aid, indices.
7) Storage:
   a) 65 to 70 degrees and 35 percent to 45 percent relative humidity.
   b) store upright on grounded metal shelves; tapes should be housed in protective fireproof cases.
   c) limit play of original; use a dubbing master to make use copies.
   d) tape life is from seven to eight years—to one hundred; varies because each tape it created differently; sensitive to environmental conditions.
   e) remove rerecord safety button and tab.

NOTES

1. Information on technical data consolidated from several sources, included the "Glossary" from *Footage 89: North American Film & Video Sources*; Brian Winston and Julia Keydal, *Working with Video: A Comprehensive Guide to the World of Video Production*, and "Guide to the Gear," *Consumer Reports* (March 1991).

2. PLANNING AND ASSESSMENT

Evaluation is very important in a videohistory shoot, during both preliminary planning and post-production. When planning, determine
whether or not video will add a visual component to your project
(whether oral history or general historical research). Evaluate the
pros and cons before you begin serious planning and scheduling.
The following questions act as a guide to evaluating and assessing
video requirements.[1]

I. Planning

1) Why do I want to use videotape?
   a) What will it add to the record?
   b) Will it complement, not duplicate, audio-taped oral history
      interviews, still photography, and other primary source
      materials?
   c) Is my purpose for using video clearly set forth? Will the
      result add to the understanding of the past? Of the inter-
      viewee(s)?
   d) Will the benefits outweigh the costs of production?

2) Will I capture unique visual data?
   a) What visual items are important to the interview?
   b) Will video help capture the person in his/her surround-
      ings in a natural way? Will it enhance or distort that re-
      lationship? Is the site necessary for the interview?
   c) Am I more interested in a "talking head" oral history in-
      terview? Is video necessary for that purpose?

3) Will video distort the interview process?
   a) What is the balance between what I hope to capture vis-
      ually and what I hope to capture intellectually (textually)?
   b) Will the presence of a technical crew alter or affect the
      interview . . . or my relationship with the interviewee?
   c) Will technical matters drive production and overwhelm
      content? How do I achieve an appropriate balance? Will
      concern for the technical aspects of the shoot affect my
      preparation as an interviewer? Do I need a new set of
      skills in dealing with technical personnel, equipment, and
      issues? Should I use a professional video producer to help
      with the shoot?

4) Have I clearly expressed the goals of the taping to:
   a) the participant? (does he/she understand legal rights? restriction possibilities, release forms?)
   b) the technical crew? (do they understand the differences between taping for historical purposes rather than for a scripted broadcast?)
   c) the institution, school, or agency for whom the interviews are conducted? (do they have special requirements or requests? are there any unusual arrangements?)

5) What do I do with the videotapes and resulting materials once taping is completed? Where will they be stored?
   a) What is required for duplicating and housing tape? Number of generational copies? Temperature and humidity?
   b) Will the tapes be transcribed? How will the video's "visual information" be noted on the printed transcript (i.e., action, process, site, appearance)?
   c) How will the material be made accessible to researchers? Finding aids? Tapes and transcripts? Equipment availability?
   d) What will we do if playback equipment changes? Do we upgrade machinery and continue to re-master tapes to fit the equipment? Is this financially feasible?

II. Assessment

1) What contribution does the product make to the historical knowledge of the subject area?
   a) Was new information was collected?
   b) Was video necessary?
   c) Did the visual dimension enhance understanding or add new knowledge?

2) Where did the video succeed—and fail?
   a) What did add to general historical knowledge?
   b) What did to add for potential public education, exhibits, etc?

3) How well did the interviewer do his/her job?
   a) Did he pursue pertinent lines of inquiry?
   b) Did he follow the "seams" or undercurrents in the interview?
   c) How focused were his questions, i.e., were his questions

concrete enough to elicit responses pertaining to visual information?

   d) Did the interviewer have knowledge of the subject? Was it apparent that he had done background research?

4) If a group interview, how well did video capture interaction?

   a) Did video capture the relationship between the participants?

   b) Were they forthcoming or reticent?

   c) Was video worth the effort in capturing this particular group on tape, for the historical record?

5) If a single-person interview, how well did video capture or convey that person's personality, knowledge, etc?

   a) Was the interview primarily a "talking head?"

   b) Was the site or an artifact incorporated into the interview effectively?

   c) Did video add anything to the interview? Or should it have been conducted on audio tape?

6) What was the technical quality of the videotaped interview?

   a) Did the lighting enhance or distract the interview subject?

   b) Were camera operations (zoom, pan, focus, etc) smooth and efficient?

   c) Were audio levels good for both interviewer and interviewee?

   d) Was the placement of all participants appropriate for the camera to capture relevant discussion?

NOTES

1. These questions were compiled from general conversation and assumptions at the SVP, from SVP committee members and participating historians, and from an excellent list of questions posed by Jeffrey Brown, professor of history at the University of New Mexico, during his comment on a panel discussion about videohistory at the joint meeting of the National Council of Public History and the Southwest Oral History Association in San Diego, March 1990. Panel members were Selma Thomas, Ava Kahn, and Terri Schorzman.

II. VIDEOHISTORY PRODUCTION

1. SMITHSONIAN VIDEOHISTORY PROGRAM

MILESTONES FOR A VIDEO PROJECT

The following outline lists milestones for projects undertaken in the Smithsonian Videohistory Program. The timetable did not include the formal proposal and review process, evaluation of the final product, nor archival processing. It does, however, represent the best estimate of the time involved in producing quality videotaped historical research.

PRE-PRODUCTION:

Twelve Weeks (minimum)

1) historian and program staff meet for preliminary planning
2) confirm potential participants and date of taping
3) select video consultant (producer/director) (if necessary)
4) submit contract for consulting services
5) submit bid requests for technical services and personnel (crew can be subcontracted through producer/director)
6) identify potential sites for taping sessions
7) schedule site inspections

Ten Weeks

1) survey site with producer
2) identify equipment and technical requirements
3) receive bids for technical services
4) prepare final budget
5) confirm location, participants, and date

Eight Weeks

1) receive bids for technical services
2) submit contracts for technical services

3) request research assistance, if needed
4) select artifacts, photos, slides
5) prepare general script/taping schedule/thematic outline
6) plan travel requirements

Four Weeks

1) arrange security and other building requirements at session location
2) make travel arrangements
3) send "participant information packet"

One Week

1) prepare site as necessary
2) final arrangements, assurances
3) travel advances (cash)

PRODUCTION:

At Session

1) meet with producer, crew, and participants for session
2) hold reception, if requested/required
3) collect signed release forms
4) review quality of tapes

POST-PRODUCTION:

Includes thank you letters and written report, as well as the preparation of name list, duplication, transcription, audit checking and copy-editing, and evaluation. Plan on at least 50 hours or more, per one hour of tape, to complete these tasks.

2. SMITHSONIAN VIDEOHISTORY PROGRAM

BUDGETING

A budget for videohistory includes costs for production, travel, and archival processing. The SVP budget did not include post-production editing. Prices for equipment and technical services varied according to production needs and regional rates. See "Average Costs."

Video Services

1) Producer:
   Planning and pre-production days, travel time, production (actual taping sessions), and "post-production" (the producer submits a report following each project, and will review the tape with the historian and staff when possible). Some producers will charge less for the non-production days.
2) Crew and Equipment:
   Price will vary according to the recording system, number of cameras, and amount of equipment used. More cameras mean more crew members, which can get expensive. A one-camera two-person professional Betacam crew averages $1,500, while a three-camera shoot in a studio may include upwards of eight people and cost over $6,000 per day.
3) Tape Stock:
   May be included in the cost of equipment and crew, although it is less expensive to purchase it in bulk and provide it for the shoot.
4) Other:
   Crew lunches and snacks.

Travel

The SVP supported travel for historians, producers, participants, and in some cases, the crew, including:

1) Transportation:
   To and from the airport (taxis, parking, etc); mode of travel to the interview site, and rental car while there.
2) Per diem:
   Hotel and meals. Government rates vary according to city.

3) Miscellaneous:
Supplies purchased on site for the shoot, photography supplies, etc.

Research

Research and production assistance is important. An assistant tracks paperwork and logistics, helps historian locate documents, artifacts, and other supporting information. The amount of help needed varied with each project, and a SVP staff member usually assisted. We occasionally hired a contractor to help with tasks. Should be included in a budget.

Archival Processing

Duplication and transcription fees multiply quickly. Check local production facilities for latest rates. The SVP paid between $35 and $50 an hour for professional dubbing from master tapes and dubbing masters, and between $75 and $95 for one hour of recorded tape for transcription services. Each stage of duplication is an additional fee:

1) Duplicate master tapes to 3/4″ dubbing master;
2) Duplicate dubbing masters to 1/2″ use copies;
3) Duplicate use copies to audio tapes for transcription (can be done during production, if sound person is willing to monitor it);
4) Transcriber costs may be higher for a group interview since the transcriber may need to view the tape to learn speakers.

Misc. Expenses

Misc. expenses might be needed for:

1) Production:
Photo enlargements, set design requirements, film for photo stills, film processing.
2) Social functions:
Reception, luncheon, etc., related to the taping sessions.
3) Research supplies:
Paper, pens, photocopying, long-distance calls; high quality audio cassette tape recorder and tapes.

3. SMITHSONIAN VIDEOHISTORY PROGRAM

AVERAGE VIDEO AND ARCHIVAL PROCESSING COSTS

The following figures, while based on actual costs, are approximate, and suggest the range of costs likely to be encountered with different types of production. Variations will depend on local market conditions as well as requirements for experience, facilities, and equipment, and special services.

Technical and Production Costs

Three camera shoot using 1″ tape $6,000.00/day plus

> Note that many video projects of this complexity require extensive set-up time (often a full day) and up to eight crew. Figure includes tape cost (open reel).

One camera shoot using Betacam
(high speed 1/2″) $1,500.00/day

> Includes a two person crew; does not include tape cost. A single, 20-minute Betacam tape costs between $10.00 and $20.00.

Costs for Directors

Average fee $400.00/day

> Fees vary depending upon experience, and in our experience have ranged from $200/day to $600/day. Some producers charge less for pre-production planning, more for actual production.

Average number of days worked   6 days/project

> Directors work between two and eleven days on a videohistory project, depending on its complexity. Work includes site survey, selection of crew, meetings, technical arrangements, consultation with historian regarding use of artifacts and sites, and purpose and goals of project.

Average total cost $2,400.00/project

Duplication Costs

The following figure assumes a four-hour duplication job, with a 1″ or Betacam master dubbed to 3/4″ U-Matic with a set of three 1/2″ VHS copies. The Videohistory Program uses time-base correction to preserve visual quality in duplication, as well as time-code windows on all VHS dubs. Duplication costs vary widely and are often negotiable.

Dub to 3/4″ U-Matic, with time-base correction $ 50.00/hour

Dub to two-hour VHS, with time-code window $ 50.00/hour

Transcription Costs

The cost for transcribing group interviews is slightly higher than a single-person interview. We used transcription services that charge a flat fee per hour of recorded tape, not for the time taken to transcribe that one hour of tape (which can be upwards of 15 hours).

Group interviews $ 95.00/hour

Single person interview $ 85.00/hour

Total Archival Processing Costs

Archival processing of videotaped interviews, in addition to duplication and transcription, includes editing, audit-checking, time-coding/cross-referencing, researching and preparing finding aids, indices, and other reference material. The figure below represents an estimated total archival processing costs for one hour of tape. The second figure averages the total time needed to process one hour of tape.

Average cost of all processing $ 700.00/hour
    (duplication, transcription, editing, etc.)

Average number of hours spent for processing,
    both contractor and staff time   45 hours/hour

Travel Costs

Travel is often required to complete a videohistory production, including transportation and per diem costs for participants, and/or staff and crew. As of mid-1990, the Program's travel costs (including overseas travel) averages $990 per person. The figure below represents the average travel cost incurred for the creation of one hour of tape.

Travel, both domestic and international $ 510.00/hour

4. SMITHSONIAN VIDEOHISTORY PROGRAM

REQUISITION FOR TECHNICAL VIDEO SERVICES

Services to be provided by:

Contractor name:
Company name:
Contractor address:
Contractor telephone:

Justification:

As necessary, including description of bids or other factors contributing to selection of contractor.

Statement of Work:

The contractor will provide [*number*] day[s] of production and technical services, including all staff and equipment, for a videohistory recording to document [*nature of project: for instance, "a panel discussion convened as part of the twenty-fifth anniversary reunion of the Mariner space exploration team."*] Videotaping is scheduled for [*date*], and will take place at [*location*].

Scope of Work:

Direction—The contractor will work under the immediate direction of a video producer [*include name*], who will oversee and approve set development and lighting design, as well as direct on-site taping live editing. The producer will be working in conjunction with a Smithsonian Institution historian [*include name*], who shall be the final authority in all matters relating to the content of the recordings.

Eight-hour day—A working day shall be defined as eight hours on-site [*may vary depending on needs of producer*], including set-up and break-down times.

Personnel—The contractor shall provide a [*# of people*] technical crew sufficiently experienced to assure a high-quality video production in conformance with the needs of the Program's participating historians. Contractor's staff shall be responsible for camera operation, audio recording, VTR operation, lighting, equipment setup, breakdown, and transportation of equipment and personnel. Technical crew shall include:

> # camera operators
> 1 technical director
> 1 audio engineer
> 1 lighting director
> grips and gaffers as necessary

Equipment—The contractor shall provide all equipment for broadcast-quality, [# of cameras], [*live mixed\ENG*] recording in the [*format name*] format, including:

> [#] cameras with tripods
> 1 video switcher
> [#] [format] videotape recorders
> 1 [format] video playback machine
> [#] microphones with stands and/or clips
> 1 audio mixer
> All necessary lighting equipment
> Cables and test equipment as necessary.
> Time code generator

Equipment Failure—in the event of equipment malfunctions or failures, all charges will be suspended during the time period commencing with the start of the equipment malfunctions and ceasing with the repair or replacement of the malfunctioning equipment.

Set and Set Construction—[varies with production]

Tape Stock—the Smithsonian Videohistory Program will provide all raw tape stock, and any unused tape shall be returned to the Smithsonian Institution.

Travel Time and Mileage—[varies depending on policy of production house—usually we allow around $.23/mile travel, not to exceed a certain limit].

Reimbursable expenses—costs for expendable supplies or incidental services not otherwise covered by this contract may be added to this contract and billed to the Smithsonian Videohistory Program at cost. All such expenditures must be approved in advance by the Videohistory Program, and supported by related documentation. Total costs for reimbursable expenses shall not exceed $100.00.

Additional Specifications:

° Time code

Recording specifications—time code shall be recorded on all tapes to reflect the total elapsed program time, starting with 00:00:00 at the beginning of the program, and running uninterrupted (except for breaks called by the participating historian) through the course of the program. Code will be generated and applied at all times when recorders are running, and will not be generated or recorded when tape is not rolling.

Inspection—At the completion of each day's recording, time code on all tapes recorded during the course of the day will be reviewed by a member or representative of the Smithsonian Videohistory Program staff to assure conformance with the above specifications.

Correction of discrepancies—Should any time-code discrepancies be noted during the above-mentioned inspection, the contractor will be responsible for their correction without additional costs to the Smithsonian Institution. Smithsonian Institution Archives specifications require that safety copies of all original tapes be made before any original is altered, so the contractor will also be responsible for the production of a complete set of U-matic safety copies before performing any necessary time-code corrections. Tape stock for these copies will be supplied by the Smithsonian Videohistory Program. Safety copies will only be required if time code corrections are needed.

° Color Bars and Tone

Standard calibration signals shall be recorded for at least one minute at the head of each tape used during the taping session (preferred), or the first tape recorded during each taping session (at minimum).

Period of Performance:

[*As necessary—usually specific days and times are provided*].

Payment Schedule:

[*As necessary—usually a lump-sum payment*]. Payment shall made upon satisfactory receipt of all recorded tapes, and submission of a proper invoice to the Smithsonian Videohistory Program office, and its acceptance by the Videohistory Program Manager.

5. SMITHSONIAN VIDEOHISTORY PROGRAM

INTRODUCTION TO TIME CODE

What Is It

Developed for use in the television industry, time code allows video users to refer to or calculate the length of specific parts of a videotape recording. While video directors use time code to identify particular scenes for editing, time-coding or archival videotapes makes it possible for researchers to find specific taped passages quickly, and to cite them with the same accuracy as pages in a book. This is possible because time code labels each television image with a unique, time-oriented "address" that indicates the elapsed playback time between a given starting point—usually the beginning of the tape—and the image on view.

How It's Used

Although time code is normally invisible to the user, it can be made visible by playing it back through a time code reader. Readers usually allow time addresses to be read either from a counter on the machine itself, or by superimposing the code readout over a television image. This televised readout appears as a small box in a corner of the screen, and shows the code as it elapses in hours, minutes, seconds, and frames counted for each second (television runs at a rate of thirty frames, or pictures, per second). A typical readout looks like this:

This address is one hour, twenty-six minutes, and fifteen seconds from the beginning of the recording, with the twenty-seventh frame on display. And if this were the starting point for a particular scene, this number could be subtracted from the address shown at the end of the scene to calculate the length of the passage.

Time code can be used to show different kinds of information. For documentation where the actual time of recording is important, the code "clock" can be synchronized with local clock time. Another method, used by many television crews, is to reset the clock to zero at the beginning of each tape, and to key the tape number into the "hours" column. This results in a readout showing the elapsed time of each tape, and allows for calculations to determine the amount of blank tape remaining. The policy of the Videohistory Program, however, is to begin each recording session at a "zero" setting, and allow the code to run uninterrupted. This results in a readout that shows the elapsed session time, which we expect will be of more use to researchers. Using this system does not impair the usefulness of the code or tapes for editing purposes.

LONGITUDINAL TIME CODE RECORDING TECHNIQUE

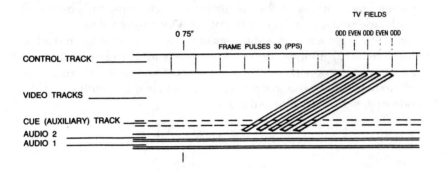

VITC RECORDING TECHNIQUE

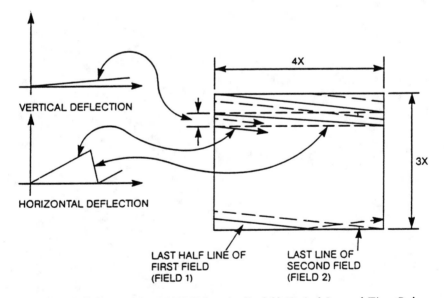

From: EECO, Inc., *SMPTE/EBU Longitudinal & Vertical Interval Time Code.*
1982.

6. SMITHSONIAN VIDEOHISTORY PROGRAM

PROJECT CHECKLIST

I. Series/Session:

Taping Date(s):

Taping Location(s):

Historian:

Producer/Director:

II. P. O. Numbers:

Producer:

Video Services:

Others:

III. To Be Completed:

___ Budget (attached)

___ Participant checklists

___ Producer/Director
___ Contract checklist completed
___ Report received

___ Historian
___ Name/word list received
___ report received
___ transcript delivered
___ transcript returned

___ Shoot information
___ Site inspection checklist
___ Video services contract checklist
___ Tapes received

__   Travel arrangements

__   Preparation for SI Archives
   __ Transcription
   __ Dubbing
   __ Other processing

IV. Production Checklist

A.   Site Inspection

   1) Site manager or representative (name and address):

   2) Lighting:

   3) Power availability:

   4) Background noise:

   5) Load-in facilities:

   6) Floor plan:

   7) Site availability:

   8) Potential annoyances:

   9) Special arrangements:

B.   Contract for Video Services

   1) Video format:

   2) Cost:

   3) Equipment:
      Lighting—
      Cameras—
      Audio—
      Time code—

    4) On-screen text:

    5) Props, furniture, or backdrops:

    6) Personnel:

C.   Producer Contract

    1) Services:

    2) Cost:

    3) Payment schedule:

    4) Address/Phone:

D.   Participant Information

    1) Participant name:

    Address:

    Telephone—Home:

            Work:

    Social Security Number:

    2) Travel Information:

    Preferred transportation:

    Nearest terminal:

    Travel information:

    Hotel information:

How to receive tickets:

    3) Travel Actions:

    SI-15 & Advance forms submitted:

    Travel Nos. to SI Travel:

    Advance & Tickets picked up:

    Voucher submitted:

4) Paperwork:

Release Form received:

Transcript sent:

Comments:

V. Processing

(see more detailed processing logs)

1) Editing:

Copy edit completed:

Sent to investigator:

Received from investigator:

Final copy:

2) Finding aid:

Project statement:

Series statement:

Preface to session:

3) Index

4) Name list

5) Photos

Materials Collated and Bound

7. SESSION CHECKLIST

Items Needed at a Videohistory Shoot

1. ___ PROJECT FILE
   ___ Release forms
   ___ Project "checklist"
   ___ Travel papers
   ___ Copy of contract
   ___ Names, addresses, phone numbers of participants
   ___ Name, address, phone number of contact/location of shoot
   ___ Name, addresses, phone numbers of contractors
   ___ Script/storyboard of session

2. ___ QUIET SIGNS

3. ___ SUPPLIES (to take, or to be supplied at site)
   ___ Gaffer's tape
   ___ Scotch tape
   ___ Scissors
   ___ Felt-tip markers (wide and narrow)
   ___ Plain paper
   ___ On-screen titles (computer generated)
   ___ Easel pad or flip chart

4. ___ EQUIPMENT (if required)
   ___ Easel
   ___ Slide projector box
   ___ Small pointer
   ___ Camera, lenses, flash, film (slide)
   ___ Triangle table
   ___ Video tape
   ___ Extension cord
   ___ Adapter

5.   __ REFRESHMENTS
        __ Soft drinks
        __ Cookies, pastries
        __ Cups
        __ Plates
        __ Napkins
        __ Pitcher of water

Notes:

Note: The SVP sent letters to participants at least two weeks before the video
    session to let them know what they might expect. We included the page
    below, as well as the "Clothing Recommendations."

8. SMITHSONIAN VIDEOHISTORY PROGRAM

PARTICIPANT INFORMATION SHEET

The Smithsonian Videohistory Program welcomes your participation in our effort to create a visual record of American scientific history. Thank you for helping assure the preservation of this heritage by contributing your time and by sharing your experiences with us. To let you know more about the way the program works, and to help you prepare for your participation, we have prepared the following information.

Background

In 1986 the Alfred P. Sloan Foundation awarded a four year grant to the Smithsonian Institution to create the Smithsonian Videohistory Program. The purpose of the Program is to encourage the exploration and use of video as a medium for documenting the history of science and technology. Under the general theme, "Science In National Life," the Program supports Smithsonian historians who wish to complement their ongoing research by creating visual records that document scientific practices and developments in their respective subject areas.

The primary goal of the Program and its participating historians is to capture information that, without a visual record, might otherwise be lost. To assure the preservation and continued availability of this record to contemporary and future scholars, all videohistory materials collected by the Program are permanently deposited in the Smithsonian Institution Archives, along with finding aids and copies of all recordings to facilitate their research use.

Travel

When participants must travel to reach a taping site, the Videohistory Program will make all travel arrangements and pay all expenses. The nature of these arrangements will vary with the circumstances, as well as the desires of the participants, but in most cases the Program will issue travel tickets approximately two weeks prior to each videohistory session, and will offer each traveler either reimbursement for expenses or a travel advance. In every case, a representative of the Program will contact the traveler to initiate travel arrangements and explain the travel program in more detail.

Release Forms

Since our video sessions are governed by copyright, we will ask you to sign copyright release forms at the time of the taping. These releases allow us to make the tapes and derivative materials available to researchers. The participating historian and a Program staff member present at your taping session will provide you with more detailed information.

For More Information

We hope your videohistory session will be both interesting and stimulating, and thank you for your contribution to the Smithsonian Institution's growing record of the history of science. A representative of the Videohistory Program will contact with you to provide specific details about your session, and will be available to answer any questions. Should you have questions or require information at any other time, please contact us at this address:

ADDRESS

9. SMITHSONIAN VIDEOHISTORY PROGRAM

CLOTHING RECOMMENDATIONS FOR VIDEOTAPED ORAL HISTORY INTERVIEWS

The Smithsonian Videohistory Program makes every effort to conduct its video sessions in a relaxed, informal atmosphere. The sessions you will be participating in are intended to gather and preserve historical information and are not theatrical productions. We do not use make-up. Due to the technical limitations imposed by video cameras and lighting, however, we suggest that you consider the following clothing recommendations when dressing for the sessions.

We suggest:

Clothing in subdued shades, in either solids or simple patterns.

Please avoid:

Large articles of brightly-colored clothing, particularly whites and reds. Bright, intense colors can overwhelm the camera.

Small and prominent patterns, since they generate distracting effects when seen on television.

Photochromatic (photogray or light sensitive) glasses, which may darken under the lights.

10. SMITHSONIAN VIDEOHISTORY PROGRAM

RELEASE FORMS: AN EXPLANATION

The Smithsonian Institution may not make your contribution to history available to research unless you sign a **release form** that transfers your share of copyright in the material to the Smithsonian Institution. A copy of the general release form is attached, and you are entitled to an explanation of its provisions and their effect.

You and the Smithsonian Institution share copyright in the material recorded on videotape—from the moment it is recorded. In order for us to make information from the videotape available to scholars, we must have from you a gift of your share of the copyright to the Smithsonian Institution.

The attached form does just that. The three paragraphs may seem redundant, but the Smithsonian General Counsel's office advises that all three paragraphs are necessary to make sure everything is clearly understood between us.

Paragraph One gives the Smithsonian your share of the copyright. **Paragraph Two** makes it clear that the physical property of the tape and any transcripts belong to the Smithsonian. **Paragraph Three** says that you place no restriction on research use of your contribution to the project.

If you have serious reservations about making any of the material available for research, please let us know. We will discuss the alternative of temporarily restricting parts of the material until your concerns are satisfied by the passage of time.

We hope that you will sign the attached unconditional release in the interests of the Smithsonian's commitment to the increase and diffusion of knowledge.

Your contribution to the growing body of research material on subjects related to "Science in National Life" is important to understanding the role of science and technology in the United States, and we look forward to sharing your experience and views to further that understanding.

11. SMITHSONIAN VIDEOHISTORY PROGRAM

RELEASE OF INTERVIEW MATERIAL

In interest and consideration of the increase and diffusion of knowledge to which the Smithsonian Institution is committed I, _____ _____, of _____ _____, hereby donate to the Smithsonian Institution any and all copyright and any and other rights, title, and interest that might exist or I may have in the interview(s) granted by me to the Smithsonian Institution on the following dates:

At the same time, I also hereby transfer and donate to the Smithsonian Institution any and all rights, title, and interest in and to any and all physical properties, including but not limited to videotapes, audiotapes, and transcripts, that fix the above-referenced interview in tangible form.

The information disclosed by me will be made available without restriction for research in accordance with the general procedures of the Smithsonian Institution Archives.

_____   _____

Date   Signature

12. SMITHSONIAN VIDEOHISTORY PROGRAM

RELEASE OF INTERVIEW MATERIAL
OPTION FOR RESTRICTION

In interest and consideration of the increase and diffusion of knowledge to which the Smithsonian Institution is committed I, _____ _____ , of _____ _____ , hereby donate to the Smithsonian Institution any and all copyright and any and other rights, title, and interest that might exist or I may have in the interview(s) granted by me to the Smithsonian Institution on the following dates:

At the same time, I also hereby transfer and donate to the Smithsonian Institution any and all rights, title, and interest in and to any and all physical properties, including but not limited to videotapes, audiotapes, and transcripts, that fix the above-referenced interview in tangible form.

The information disclosed by me will be made available without restriction for research in accordance with the general procedures of the Smithsonian Institution Archives, except for passages restricted as follows (attach additional pages as necessary):

Passages so marked and initialed will not be made available to the public until _____ (date), after which they will be made available without reservation.

Other comments:

_____   _____
Date   Signature

13. SMITHSONIAN
VIDEOHISTORY PROGRAM

FIELD FOOTAGE LOG

Project: _____   Page ___ of ___

Session: _____   Date _____

Tape #	Timecode	Description/Question/Topic	Visuals

14. SET DESIGN FOR
VIDEOTAPED GROUP INTERVIEWS

The Smithsonian Videohistory Program used a variety of arrangements for interviews with groups of three or more people—with one, two, and three cameras. Designs below include those that helped generate discussion and facilitate the use of artifacts, as well as those that did not.

Inverted "V"

Good eye-contact for participants (A–D). Historian (E) sat at open end of "V" but remained on camera, resulting in his backside being visible during group shots, which was visually distracting. Three cameras provided closeups, two-shots, and group shots. Historian used easel to display visual aids.

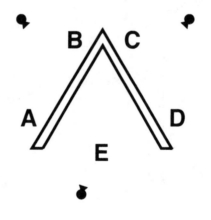

◀ = Camera Placement

Wide Inverted "V"

I. Participants (B–D) were too spread out for face-to-face interaction. Their microphones were taped down behind the table, which hindered free use of objects, which were placed on a low table in front. Historians (A & E) were on camera at all times. Had feeling of a government "briefing room." Three cameras, flexible angles, good close-ups.

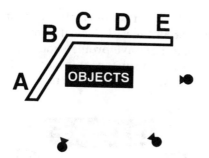

II. Participants (A–F) sat along side tables, co-moderator/participant (G) was at apex, and historians (H & I) were off-camera, at each end of table. Placed most emphasis on (G). Three cameras provided good group shots, two shots, and close-ups.

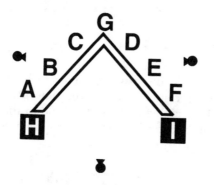

Wide "U"

I. Participants (A–C, E, F) were too spread out for interaction; historian (D) had trouble seeing everyone with ease, and his placement made him the focal point. Two cameras, with site constraints, resulted in only profiles of (A & F).

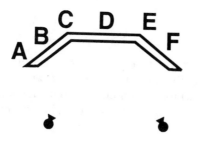

II. Participants (A–G) sat on comfortable chairs and sofas, historian (H) was off camera at all times. Informal atmosphere. Two cameras were arranged so that extensive profile shots were avoided. Artifact (X) was placed nearby, and since people were miked with lavalieres, they could spontaneously move to artifact.

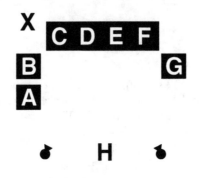

Geometric "U"

Participants (A–D) sat on swivel chairs, which was a mistake, and historian (E) was on camera most of the time. Participants deferred to mentor (A). Low coffee table added little to the arrangement. A regular table would have been more appropriate. Shot with two stationary cameras.

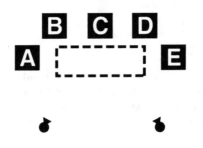

Semi-Circle

Large group (A–E) was placed in wide circle on a platform without a table. Historian (F) was off camera—on an extended platform— with side table (xxx) for documents. Two cameras were placed to the left of the historian, and one to the right, for maximum shots of participants.

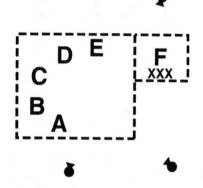

Linear

I. Full body exposure for participants (A & C) and historian (B); shot with one camera (caused side view of two faces). Uncomfortable and constrained. Large artifact (xxxx) behind group, and lack of close-ups, dwarfed participants.

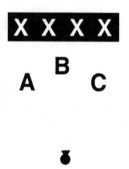

II. Participants (A–D) were lined in row, without a table. Group was obviously uncomfortable with design and had no face-to-face interaction. Historian (E) was off-camera. Shot with one camera, which captured both close-ups and wide angles due to attentive camera operator.

III. Participants (A–D) sat at table; historian (E) sat off camera, between two cameras. Another table, to right of participants, held artifacts. An assistant passed them to participants when needed. Two cameras captured all shots, including reaction and close-ups. Not much "face-to-face" interaction due to linear arrangement.

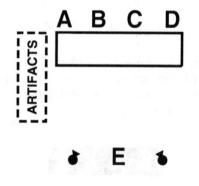

Round Table

Better design for participant's (A & B) interaction, but awkward positioning. Provided table space for artifacts and notes. Historian (C) was on camera at all times.

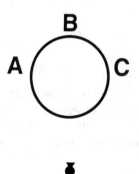

Triangle Table

Arrangement brought participants (A–D) and historian (E) closer together and facilitated discussion. Encouraged use of artifacts (placed on table). Three cameras allowed for full-face views and close-ups of people and artifacts, as well as of the interviewer. Only two cameras are necessary for this arrangement to adequately capture the interviewees. Used arrangement on several occasions with success.

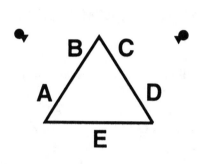

III. VIDEOHISTORY POST-PRODUCTION

1. SMITHSONIAN VIDEOHISTORY PROGRAM

ARCHIVAL REQUIREMENTS SUMMARY

I. Tape Labels

1. Labels, for all tape generations, should include:

Information	Example
NAME OF PROGRAM	SMITHSONIAN VIDEOHISTORY PROGRAM
Name of Project	Multiple Mirror Telescope
Name of Historian	David DeVorkin, NASM
Session Number, Date	S-3, May 8, 1989
Description	Engineering/design; Group
Location	Tucson, Arizona
Length	45 minutes
Generation	**1st generation Master**
Number in sequence	Tape 2 of 2

2. Labels, for each generation, should be a different color.
   A) Yellow—first generation, master
   B) Blue—second generation, dubbing master
   C) White—third generation, reference copy

II. Transcripts

Videohistory transcripts contain more than oral narrative. A good working transcript should contain information such as time-code cross references and visual cues.

1. *Time Code.* Place time codes on the transcript when the subject changes, when an artifact is introduced, or every couple of

minutes. Time code will help the researcher find segments of tape for viewing, rather than watching the entire (if time is limited).

01:33:05:00
SMITH: Please show me how the vacuum testing chamber works.

JONES: O.K. First, you lift the lever, then. . . . .

03:20:00:00
SMITH: Where did you learn to do that?

JONES: Oh, I trained here twenty years ago with. . . .

2. *Visual Cue.* Visual cues are short, direct, active statements about what occurs visually on tape. This information helps a researcher understand the action, if not made clear in the narrative.

02:45:22:00   [Camera follows Genin through laboratory and specimen collections.]

LEWIS: Dr. Genin, where are we now? Will you please describe it?
                    [Genin points to cages. Camera pans the room and then shows close-up of a monkey.]

GENIN: We are in the monkey chamber of the medical institute; this is where we train monkeys to go into space.

3. *Location Changes.* A single videohistory session may take place in several locations. Note it in one of two ways:
   A) At the top center of the page. Use this approach when there are several distinct locations or "set ups."

LOCATION TWO:
MMT: Telescope Operations Chamber
Fred Chaffee, Craig Foltz, Carol Heller

   B) Within a visual cue. Use this approach when there are many locations within an interview, such as a lengthy tour of a building or manufacturing area.

00:24:30:00   [Ledley moves down hall to room that contains cat-scan prototype, Room 5-B.]

LEDLEY: I'll show you the prototype. It's in here. I designed this in 1974 . . . it's a real beauty! It took a lot of calculations, a lot of design changes . . .

4. *Visual descriptions without narrative.* Since video records visual information, there may be occasions where narrative is absent. Such information can be placed in written form.

00:01:05:00   [Conveyor carries large pieces of slate from quarry to cutting room.]

00:05:02:00   [Beayon slices one-inch thick pieces of slate from large block, tosses inferior pieces into scrap pile, and sends quality slate down conveyor for precision cutting].

5. *Two-cameras, separate recording.* When two cameras have simultaneously recorded an interview, images from both cameras should be noted. The first cue is usually camera A, the second is camera B.

02:00:35:00   [A: Comstock nods head in agreement with Carlson.]
[B: K. Williams and R. Williams both frown and shake heads in disagreement].

CARLSON: The business did really well five years ago. It's only recently that we've had organizational problems. Then came design problems, then a general financial downturn in the industry. We're coming back, though.

2. SAMPLE FINDING AID:
VERMONT STRUCTURAL SLATE

SMITHSONIAN INSTITUTION ARCHIVES   p. 1
SMITHSONIAN VIDEOHISTORY PROGRAM
October 1989   9547

Vermont Structural Slate

The Vermont Structural Slate Company (VSS) in Fair Haven, Vermont, was founded in 1859. As of 1989, it was one of twenty remaining companies nationwide that produced slate. The company operated several quarries, including the oldest active quarry in Vermont, the Eureka quarry, which opened for slate production in 1852. VSS employees have continued to use nineteenth century machinery for most quarrying and manufacturing operations. The owners, however, have attempted to upgrade the facility with more modern equipment.

William Worthington, museum specialist at the National Museum of American History (NMAH), recorded quarrying techniques at VSS on October 12 and 13, 1989, to document remaining nineteenth century industrial techniques before the installation of modern equipment. For example, Worthington recorded the operation of the old cableway system that removed slate from the Eureka pit, as well as the more modern use of cranes, diesel shovels, and dump trucks. Worthington also documented various methods and equipment used in making slate shingles.

The videohistory shoot included both interviews with employees and detailed visual documentation of their work, as well as overall tours of the quarry, its operation, and its environs. **Brad Bauman,** chief engineer, guided Worthington around the site and explained various processes involved in slate manufacture, while **Everett Beayon**, the last employee familiar with the cableway system, returned from his retirement to demonstrate and explain the operation of the system. **Joseph Root** described selecting and extracting slate from the quarry, and **Raymond Cull** demonstrated the signaling system used to communicate with crane operators for the removal of slate from the pit. A number of other employees appeared throughout both sessions, but were not interviewed.

Sessions were recorded on Betacam, and the original tapes are in preservation storage. VHS copies and full time-coded transcripts with abstracts of visual information are available for research.

Box 1   Transcripts and Videotapes of Sessions

Session One (October 12, 1989), at the Eureka Quarry of Vermont Structural Slate Company, documented quarrying operations and equipment, with an emphasis on the older cableway transportation system, c. 1859–1989, including:

- installation and function of "dead log" anchors for cableway systems;
- overview of the Eureka quarry pit;
- "shelf technique" method for slate removal from quarry walls;
- cableway system for lifting rocks out of the quarry;
- use of electromagnetic signaling system for communications between quarry floor workers and engine house or crane operators;
- dumping operation of rubbish box;
- control of the cableway operations from an engine house.

Visual documentation included:

- features of the cableway carriage;
- specialized hardware designed for attaching the slate slabs and rubbish box to cranes for removal from the quarry floor;
- engine house equipment and furnishings.

Transcript pages 1–42
Tape One—120 minutes
Tape Two—20 minutes

Session Two (October 13, 1989), in the roofing shed and at a rubbish heap at the Vermont Structural Slate Company, documented the process of making slate roofing shingles, c. 1989, including:

- sawing, splitting, trimming and hole punching of shingles;
- disposal of waste created in trimming process;
- preparation ("plugging") of large slate slabs for placement on the roofing shed conveyor;

- panorama of VSS grounds with narration;
- pick-up shots with narration of the "stick" support and the terminus for the cableway system.

Visual documentation included:

- step-by-step tour of all operations in the shingle making process;
- pallets of imported Spanish slate sold by VSS;
- overview of other VSS workshops and the area surrounding the Eureka quarry;
- the "stick" cableway support and the terminus slate rubbish heap.

Transcript pages 1–21
Tape One—100 minutes

Joan Mathys
August 1991

3. SAMPLE TRANSCRIPT: FIRST PAGE FROM VERMONT STRUCTURAL SLATE

1

00:01:21:00	[Begin Session One] [Begin VHS Tape 1 of 2] [Begin U-Matic Tape 1 of 3]
00:01:29:00	[Worthington and Beayon stand near the eastern end of the quarry]

WORTHINGTON: I'm William Worthington from the Division of Engineering and Industry at the National Museum of American History. Today, October 12, 1989, we are at the Eureka quarry of the Vermont Structural Slate Company at Fairhaven, Vermont.

We are going to record the operation of one of the remaining cable hoist systems that was used to remove slate from the quarry. Can you tell us who you are and what your position was?

BEAYON: I'm Everett Beayon. I was a rigger here. Sam Jones used to be the rigger before me, and I learned from him how to do this.

00:02:11:00 [Worthington indicates pulley attachment leading into rock pile]

WORTHINGTON: Can you tell us what this is?

BEAYON: This is where the main cable is hooked into the deadlog.

WORTHINGTON: What's a deadlog?

BEAYON: That's what anchors the main cable so it can't pull out.

WORTHINGTON: How is it held in there?

BEAYON: Well, we have a big oak log buried behind this heap of rubbish and rock on top of it. Most of the rock is forward of the log to keep the weight on it, so it can't pull out.

4. SAMPLE TRANSCRIPT: DESCRIPTION OF NONVERBAL, VISUAL INFORMATION FROM VERMONT STRUCTURAL SLATE, PAGE 12

12

have four pins to take all the pressure that you would have on a deadlog where there is no convenient place to put a deadlog.

00:32:54:00	[Pause in tape]
00:33:13:00	[Close-up of rubbish box attached to cableway crane as it is hoisted to top of rubbish pile]
00:34:40:00	[Lowering of rubbish box]
00:35:00:00	[Close-up of Cull on a quarry bench using bell wires to signal the cableway crane operator in the engine house]
00:35:54:00	[Cull and co-worker heap waste slate into rubbish box]
00:37:26:00	[Close-up of worker attaching chains from the cableway crane to the rubbish box]
00:38:35:00	[Close-up of Cull rubbing the bell wires together to signal the operator in the engine house]
00:39:33:00	[Rubbish box is lifted out of quarry]
00:43:20:00	[Rubbish box is lowered to quarry bench again]
00:43:45:00	[Worker removes chains attaching the rubbish box to the crane]
00:44:20:00	[Cull hooks a chain around a large piece of waste slate]
00:45:38:00	[Worker brings crane over to Cull]

BIBLIOGRAPHY

BOOKS

Barnouw, Eric. *Documentary: A History of Non-Fiction Film.* Oxford, 1983.

Carson, Barbara, and Gary Carson. "Social History From Artifacts." *Ordinary People and Everyday Life: Perspectives on the New Social History*, ed. James B. Gardner and George Rollie Adams. Nashville, 1983.

Collier, Malcom, and John Collier, Jr. *Visual Anthropology: Photography as a Research Method.* New Mexico, 1986 and 1990.

Committee on Preservation of Historical Materials. *Preservation of Historical Records.* Committee on Preservation of Historical Materials of the National Materials Advisory Board, Commission on Engineering and Technical Systems of the National Research Council, National Academy Press. Washington, D.C., 1986.

Compesi, Ronald, and Ronald E. Sherriffs. *Small Format Television Production.* Boston, 1985.

Davis, Nuel Pharr. *Lawrence and Oppenheimer.* New York, 1968.

DeVorkin, David. *Science with a Vengeance: The Military Origins of the Space Sciences in the V-2 Era.* Springer-Verlag, forthcoming.

Empsucha, Joseph P. "Film and Videotape Preservation Factsheet." *Footage 89: North American Film and Video Sources*, ed. Richard Prelinger. New York, 1989.

Frisch, Michael. *A Shared Authority: Essays on the Craft and Meaning of Oral and Public History.* Albany, N.Y., 1990.

Herman, Gerald H. "Media and History." *The Craft of Public History: An Annotated Select Bibliography*, ed. David F. Trask and Robert W. Pomeroy III. Westport, Conn., 1983.

Hewlett, Richard, and Oscar E. Anderson, Jr. *The New World 1939–1946.* Pennsylvania State University, 1962.

Hoke, Donald. *Ingenious Yankees: The Rise of the American System of Manufacturing in the Private Sector.* New York, 1990.

Jackson, Bruce. *Fieldwork.* Urbana, 1987.

Jolly, Brad. *Videotaping Local History.* Nashville, 1983.

Kennedy, George P. *Vengeance Weapon 2: The V-2 Guided Missile.* Washington, D.C., 1983.

225

McMahan, Eva M. *Elite Oral History Discourse: A Study of Cooperation and Coherence.* Tuscaloosa, 1990.

Mead, Margaret. "Visual Anthropology in a Discipline of Words." *Principles of Visual Anthropology,* ed. Paul Hockings. The Hague and Paris, 1975.

Moore, Charles W. *Timing a Century.* Cambridge, Mass., 1945.

Newell, Homer. *Beyond the Atmosphere.* NASA, 1980.

Neuenschwander, John N. *Oral History and the Law.* Denton, Tex., 1985.

O'Connor, John. *Image as Artifact: Historical Analysis of Film and Television.* Malabar, Fla., 1990.

Ordway, Frederick I., and Mitchell R. Sharpe. *The Rocket Team.* New York, 1979.

Perliss, Vivian. "Oral History A Bibliography." *The Past Meets and Present: Essays on Oral History,* ed. David Stricklin and Rebecca Sharpless. Lanham, Md., 1988.

Prelinger, Rick. *Footage 89: North American Film and Video Sources.* New York, 1989.

Rhodes, Richard. *The Making of the Atomic Bomb.* New York, 1968.

Schlereth, Thomas J. *Material Culture Studies in America.* Knoxville, 1982.

———. "Material Culture and Cultural Research." *Material Culture: A Research Guide,* ed. Thomas J. Schlereth. Lawrence, Kans., 1985.

———. *Cultural History and Material Culture: Everyday Life, Landscapes, Museums.* Ann Arbor, 1990.

Schorzman, Terri. "The Library as Producer: The Creation of Local and Personal Video Collections." *Video Collection Management and Development* (Greenwood Publishing Group), forthcoming.

Vincenti, Walter G. *What Engineers Know and How They Know It: Analytical Studies from Aeronautical History.* Baltimore, 1990.

Winston, Brian, and Julia Keydel. *Working with Video: A Comprehensive Guide to the World of Video Production.* London, 1986.

Zettle, Herbert. *Television Production Handbook.* Belmont, Calif., 1984.

JOURNALS

Azarm, Shapour, Jingen Chen, and L. W. Tsai. "Walking Robot: A Multidisciplinary Design Project for Undergraduate Students." *International Journal of Mechanical Engineering Education* 18 (April 1990).

Bares, J. "Ambler: An Autonomous Rover for Planetary Exploration." *IEEE Computer* (June 1989): 18–26.

Carosso, Vincent P. "The Waltham Watch Company: A Case History." *Bulletin of the Business Historical Society* 23 (December 1949): 179.

Charlton, Thomas. "Videotaping Oral Histories: Problems and Prospects." *American Archivist* 47 (Summer 1984): 228–236.

Chepesiuk, Ron, and Ann Y. Evans. "Videotaping History: The Winthrop College Archives' Experience." *American Archivist* 48 (Winter 1985): 65–68.

Consumer Reports. "Guide to the Gear." (March 1991).

Diament, Sarah E. "Can Film Complement Oral History Interviews?" *The Fourth National Colloquium on Oral History* (1970): 122–27.

Frantz, Joe B. "Videotaping Notable Historians." *The Third National Colloquium on Oral History* (1969): 89–101.

Gardner, Joel. "Oral History and Video in Theory and Practice." *Oral History Review* 12 (1984): 105–111.

Goldberg, Stanley. "Inventing a Climate of Opinion: Vannevar Bush and the Decision to Build the Bomb." *Isis* (forthcoming).

Handfield, F. Gerald, Jr. "The Importance of Video History in Libraries." *Drexel University Quarterly.* 15 (October 1979): 29–34.

Henson, Pamela, and Terri Schorzman. "Videohistory: Focusing on the American Past." *Journal of American History* 78 (September 1991): 618–27.

Hindle, Brooke. "Presidential Address: Technology Through the 3-D Time Warp." *Technology and Culture* 24 (1983): 452.

———. *Technology in Early America: Needs and Opportunities.* Chapel Hill, 1966.

Jameson, Elizabeth, and David Lenfest. "From Oral to Visual: A First-Timer's Introduction to Media Production." *Frontiers* VII (1983): 25–31.

McCurdy, Howard. "The Decay of NASA's Technical Culture." *Space Policy* 5 (1989): 301–10.

Michigan Oral History Council. "Public Access Television Opens New Era in Oral History." Michigan Oral History Council (Winter 1989): 2, 7.

Mould, David. "Composing Visual Images for the Oral History Interview." *International Journal of Oral History* 7 (November 1986): 198–205.

Murray, James Briggs. "Oral History/Video Documentation at the Schomburg Center, More Than Just 'Talking Heads.' " *Film Library Quarterly* 15 (1982): 3–7.

Oral History Association. "Evaluation Guidelines." *Oral History Association.*

Raymond, Chris. "Increasing Use of Film by Visually Oriented Anthropologists Stirs Debate Over Ways Scholars Describe Other

Cultures." *The Chronicle of Higher Education* (March 1991): A5, A8–9.

Rinzler, Ralph, and Robert Sayers. "The Meaders Family, North Georgia Potters." *Smithsonian Folklife Studies* 1 (1980).

Schorzman, Terri. "Smithsonian Videohistory Program Symposium, Conference Report." *Technology and Culture* 30 (January 1989): 118–22.

Shores, Louis. "Directions for Oral History." *Oral History At Arrowhead: The Proceedings of the First National Colloquium on Oral History* (1967): 53–66.

Sipe, Dan. "The Future of Oral History & Moving Images." *Oral History Review* 19/1–2 (Spring-Fall 1991): 75–87.

Sorenson, Richard. "A Research Film Program in the Study of Changing Man." *Current Anthropology* 8 (1967): 443–69.

Taylor, Daniel, and Mark Rawitsch. "Beyond the Cards: An Introduction to Documenting Historical Collections with Video Tape." *History News* 44 (March/April 1989).

Toplin, Robert Brent. "The Filmmaker as Historian." *American Historical Review* 93 (December 1988): 1210–27.

Walkowitz, Daniel. "Visual History: The Craft of the Historian-Filmmaker." *The Public Historian* 7 (Winter 1985): 53–64.

Walters, Ronald. "Fixing the Image." *The Public Historian* 13 (Spring 1991): 101–06.

Washburn, Wilcomb. "Material Culture and Public History: Maturing Together?" *The Public Historian* 13 (Spring 1991): 53–60.

Weiner, Charles I. "Oral History of Science: A Mushrooming Cloud." *Journal of American History* 75 (September 1988): 548–59.

Wheeler, Jim. "Long-Term Storage of Videotape." *SMPTE Journal* (1983): 650–54.

White, Hayden. "AHR Forum: Historiography and Historiophoty." *American Historical Review* 93 (December 1988): 1193–1199.

Whittaker, W. Richard. "Why Not Try Videotaping Oral History." *Oral History Review* 9 (1981): 115–124.

Wilson, David L. "Governors Past: A Video History Project." *Journal of Instructional Media* 6 (1978–79): 253–264.

Wise, George. "Science and Technology." *Osiris* 1 (1985): 229–46.

NEWSLETTERS, ARTICLES, REPORTS, SPEECHES

Alfred P. Sloan Foundation. *1982 Annual Report* (New York).
———. *1983 Annual Report* (New York).

————. *1984 Annual Report* (New York).

Annenburg Washington Program. "A Roundtable on Television Preservation." Rapporteur Summary of a Colloquium Convened by The Annenberg Washington Program. Washington, D.C. May 19, 1989.

Boyle, Dierdre. "Video Preservation: Insuring the Future of the Past." *The Independent* (October 1991): 25–31.

Brown, Jeffrey. Summary Comments at the National Council of Public History and Southwest Oral History Association jointly held meeting. San Diego. March 1990.

Conservation Administration. "Film/Videotape Factsheet." *Conservation Administration News* 22 (no date).

Hayman, Randy. "Archiving Videotape." *Audio Visual Communications* (March 1991): 21–25.

Henson, Pamela. "Visual Documentation and Historical Research." Paper presented at the Organization of American Historians and Society for History in the Federal Government jointly held annual meeting. Washington, D.C. March 1990.

Kemp, Roma. "How Long is Forever? and New Methods of Data Storage." *Newsletter, Northwest Oral History Association* (Fall/Winter 1990): 2, 4–7.

Mathys, Joan. Seminar presentation on archival processing of videotape. The National Library of New Zealand. Wellington, New Zealand. January 1991.

Oral History Association. "Oral History of and for the Deaf Begins at Gallaudet." *Oral History Association Newsletter* (Fall 1977).

Page, Don. "A Visual Dimension to Oral History." Canadian Oral History Association, *Journal* 2 (1976–77): 20–23.

Prelinger, Rick. "Archival Survival: The Fundamentals of Using Film Archives and Stock Footage Libraries." *The Independent* (October 1991): 20–24.

Schorzman, Terri. "Creating Images: The Smithsonian's Approach to Videohistory." *PHS Network* 16 (Santa Barbara, Spring, 1988).

Schorzman, Terri. "Video Plays a Role in Historical Documentation at the Smithsonian." National Council on Public History, *Public History News* (Winter 1988/89).

Schulman, David. "Deja Vu—Resorting and Remastering Open-Reel and Storage of Magnetic Recording Tape." *The Independent* (October 1991): 32–35.

————. "The Handling and Storage of Magnetic Recording Tape." *Retentivity* (no date).

———. "Videocassette Tape Physical Damage." *Retentivity* (no date).
Tomlison, Don E., and Ray Fielding. Seminar discussion, "The Computer Manipulation and Creation of Video and Audio: Assessing the Downside." The Annenberg Washington Program, Northwestern University. December 10, 1991.
Walters, Ronald. Summary Comments at the Organization of American Historians and the Society for History in the Federal Government jointly held annual meeting, Washington, D.C. March 1990.

NONPUBLISHED, ARCHIVAL, AND REFERENCE MATERIALS

DeVorkin, David. Memo to John Fleckner, Marc Pachter, William Moss, and Nathan Reingold. November 18, 1985. Smithsonian Institution Archives.
———. Memo to Video History Committee Participants. January 3, 1986. Smithsonian Institution Archives.
———. Memos to "Distribution." March 3, 1986, and March 27, 1986. Smithsonian Institution Archives.
———. Sloan Videohistory Report #1, October 19, 1986. Smithsonian Institution Archives.
———. Review of 25th Anniversary of Mariner 2 videohistory project. Spring 1988. Smithsonian Institution Archives.
Harahan, Joseph P. Report on the evaluation of videotape of 20th Century Small Arms with Eugene Stoner. November 16, 1988. Smithsonian Institution Archives.
Henson, Pamela. "Oral History Project Manual." Smithsonian Institution Archives.
Homiak, Jake. Memo to Pamela Henson. March 18, 1991. Smithsonian Institution Archives.
Kennedy, Roger. Memo to Robert McC. Adams. July 3, 1985. Smithsonian Institution Archives.
Leslie, Stewart. Report on the Rand videohistory project. 1988. Smithsonian Institution Archives.
Lewis, Alan. Fact Sheet on Videotape Preservation. January 1991.
Moss, William. Memo to Robert McC. Adams. June 21, 1985. Smithsonian Institution Archives.
———. Memo to David DeVorkin. December 4, 1986. Smithsonian Institution Archives.
———. Memo to David DeVorkin, Nathan Reingold, Marc Pachter,

and Pamela Henson. February 13, 1987. Smithsonian Institution Archives.

Meyerott, Roland E. Letter to Marcus O'Day, January 13, 1947. Lyman Spitzer Papers, Princeton University.

Reingold, Nathan. Memo to Robert McC. Adams. June 20, 1985. Smithsonian Institution Archives.

Schorzman, Terri. Status Report to Robert Hoffmann. July 12, 1990. Smithsonian Institution Archives.

Smithsonian Institution. "Proposal to the Alfred S. Sloan Foundation." June 1986. Smithsonian Institution Archives.

Smithsonian Videohistory Program. *Guide to the Collections of the Smithsonian Videohistory Program.* 1992.

———. "Annual Report: Summary of First Year Activities and Request for Second Year Funding (FY88)." September 1987. Smithsonian Institution Archives.

———. "Annual Report for Second Year Activities and Request for Third Year Funding." September 1988. Smithsonian Institution Archives.

———. Responses to Questionnaire by historians. In project files. Smithsonian Institution Archives.

Thomas, Selma. Report on 20th Century Small Arms videohistory project. May 1988. Smithsonian Institution Archives.

———. Report on the Waltham Clock Company videohistory project. July 27, 1989. Smithsonian Institution Archives.

Williams, Brien. Report on Paleontology videohistory project. June 1987. Smithsonian Institution Archives.

———. Report on Classical Observing Methods videohistory project. April 1988. Smithsonian Institution Archives.

———. Report on Multiple Mirror Telescope videohistory project. July 3, 1989. Smithsonian Institution Archives.

Williams, Michael. Report on the ENIAC videohistory project. February 1989. Smithsonian Institution Archives.

White, Stephen. Memo to Arthur Singer, October 6, 1985. Smithsonian Institution Archives.

AUDIO AND VIDEO MATERIALS

Smithsonian Videohistory Program. Videohistory tapes and transcripts for twenty-two video projects. Smithsonian Institution Archives. 1986–1992.

Taylor, Lauriston. *Vignettes of Early Radiation Workers.* 8 videocassettes. Rockville, MD., 1978.

INDEX

Page numbers for illustrations appear in boldface.